Crop Productivity

BOOKS IN THE BIOTOL SERIES

BIOTECHNOLOGY BY OPEN LEARNING

Crop Productivity

PUBLISHED ON BEHALF OF :

Open universiteit   and   **University of Greenwich (formerly Thames Polytechnic)**

Valkenburgerweg 167
6401 DL Heerlen
Nederland

Avery Hill Road
Eltham, London SE9 2HB
United Kingdom

Butterworth-Heinemann Ltd
Linacre House, Jordan Hill, Oxford OX2 8DP

ℛ A member of the Reed Elsevier plc group

OXFORD LONDON BOSTON
MUNICH NEW DELHI SINGAPORE SYDNEY
TOKYO TORONTO WELLINGTON

First published 1994

British Library Cataloguing in Publication Data
A catalogue record for this book is
available from the British Library

Library of Congress Cataloguing in Publication Data
A catalogue record for this book is
available from the Library of Congress

ISBN 0 7506 0562 6

Composition by University of Greenwich
(formerly Thames Polytechnic)
Printed and Bound in Great Britain

The Biotol Project

The BIOTOL team

OPEN UNIVERSITEIT, THE NETHERLANDS
Prof M. C. E. van Dam-Mieras
Prof W. H. de Jeu
Prof J. de Vries

UNIVERSITY OF GREENWICH (FORMERLY THAMES POLYTECHNIC), UK
Prof B. R. Currell
Dr J. W. James
Dr C. K. Leach
Mr R. A. Patmore

This series of books has been developed through a collaboration between the Open universiteit of the Netherlands and University of Greenwich (formerly Thames Polytechnic) to provide a whole library of advanced level flexible learning materials including books, computer and video programmes. The series will be of particular value to those working in the chemical, pharmaceutical, health care, food and drinks, agriculture, and environmental, manufacturing and service industries. These industries will be increasingly faced with training problems as the use of biologically based techniques replaces or enhances chemical ones or indeed allows the development of products previously impossible.

The BIOTOL books may be studied privately, but specifically they provide a cost-effective major resource for in-house company training and are the basis for a wider range of courses (open, distance or traditional) from universities which, with practical and tutorial support, lead to recognised qualifications. There is a developing network of institutions throughout Europe to offer tutorial and practical support and courses based on BIOTOL both for those newly entering the field of biotechnology and for graduates looking for more advanced training. BIOTOL is for any one wishing to know about and use the principles and techniques of modern biotechnology whether they are technicians needing further education, new graduates wishing to extend their knowledge, mature staff faced with changing work or a new career, managers unfamiliar with the new technology or those returning to work after a career break.

Our learning texts, written in an informal and friendly style, embody the best characteristics of both open and distance learning to provide a flexible resource for individuals, training organisations, polytechnics and universities, and professional bodies. The content of each book has been carefully worked out between teachers and industry to lead students through a programme of work so that they may achieve clearly stated learning objectives. There are activities and exercises throughout the books, and self assessment questions that allow students to check their own progress and receive any necessary remedial help.

The books, within the series, are modular allowing students to select their own entry point depending on their knowledge and previous experience. These texts therefore remove the necessity for students to attend institution based lectures at specific times and places, bringing a new freedom to study their chosen subject at the time they need and a pace and place to suit them. This same freedom is highly beneficial to industry since staff can receive training without spending significant periods away from the workplace attending lectures and courses, and without altering work patterns.

BIOcalm

SOFTWARE IN THE BIOTOL SERIES

BIOcalm interactive computer programmes provide experience in decision making in many of the techniques used in Biotechnology. They simulate the practical problems and decisions that need to be addressed in planning, setting up and carrying out research or development experiments and production processes. Each programme has an extensive library including basic concepts, experimental techniques, data and units. Also included with each programme are the relevant BIOTOL books which cover the necessary theoretical background.

The programmes and supporting BIOTOL books are listed below.

Isolation and Growth of Micro-organisms
Book: *In vitro* Cultivation of Micro-organisms
         Energy Sources for Cells

Elucidation and Manipulation of Metabolic Pathways
Books: *In vitro* Cultivation of Micro-organisms
         Energy Sources for Cells

Gene Isolation and Characterisation
Books: Techniques for Engineering Genes
         Strategies for Engineering Organisms

Applications of Genetic Manipulation
Books: Techniques for Engineering Genes
         Strategies for Engineering Organisms

Extraction, Purification and Characterisation of an Enzyme
Books: Analysis of Amino Acids, Proteins and Nucleic Acids
         Techniques used in Bioproduct Analysis

Enzyme Engineering
Books: Principles of Enzymology for Technological Applications
         Molecular Fabric of Cells

Bioprocess Technology
Books: Bioreactor Design and Product Yield
         Product Recovery in Bioprocess Technology
         Bioprocess Technology: Modelling and Transport Phenomena
         Operational Modes of Bioreactors

Further information: Greenwich University Press,
University of Greenwich, Avery Hill Road, London, SE9 2HB.

Contributors

AUTHORS

Dr N.J.W. Clipson, University of Wolverhampton, Wolverhampton, UK

Dr S. J. Edwards, University of Liverpool, Liverpool L69 3BX, UK

Dr J.F. Hall, De Montfort University, Leicester, UK

Dr C.K. Leach, De Montfort University, Leicester UK

Dr F.W. Rayns, The Henry Doubleday Research Association, Coventry, UK

Dr G.D. Weston, De Montfort University, Leicester, UK

EDITOR

Dr G.D. Weston, De Montfort University, Leicester, UK

SCIENTIFIC AND COURSE ADVISORS

Prof M.C.E. van Dam-Mieras, Open universiteit, Heerlen, The Netherlands

Dr C.K. Leach, De Montfort University, Leicester, UK

ACKNOWLEDGEMENTS

Grateful thanks are extended, not only to the authors, editors and course advisors, but to all those who have contributed to the development and production of this book. They include Mrs A. Allwright, Miss K. Brown, Ms H. Leather, Miss J. Skelton, and Dr A. McCormack.

The development of this BIOTOL text has been funded by **COMETT, The European Community Action Programme for Education and Training for Technology**. Additional support was received from the Open universiteit of The Netherlands and from the University of Greenwich (formerly Thames Polytechnic).

Contents

How to use an open learning text

An open learning text presents to you a very carefully thought out programme of study to achieve stated learning objectives, just as a lecturer does. Rather than just listening to a lecture once, and trying to make notes at the same time, you can with a BIOTOL text study it at your own pace, go back over bits you are unsure about and study wherever you choose. Of great importance are the self assessment questions (SAQs) which challenge your understanding and progress and the responses which provide some help if you have had difficulty. These SAQs are carefully thought out to check that you are indeed achieving the set objectives and therefore are a very important part of your study. Every so often in the text you will find the symbol Π, our open door to learning, which indicates an activity for you to do. You will probably find that this participation is a great help to learning so it is important not to skip it.

Whilst you can, as an open learner, study where and when you want, do try to find a place where you can work without disturbance. Most students aim to study a certain number of hours each day or each weekend. If you decide to study for several hours at once, take short breaks of five to ten minutes regularly as it helps to maintain a higher level of overall concentration.

Before you begin a detailed reading of the text, familiarise yourself with the general layout of the material. Have a look at the contents of the various chapters and flip through the pages to get a general impression of the way the subject is dealt with. Forget the old taboo of not writing in books. There is room for your comments, notes and answers; use it and make the book your own personal study record for future revision and reference.

At intervals you will find a summary and list of objectives. The summary will emphasise the important points covered by the material that you have read and the objectives will give you a check list of the things you should then be able to achieve. There are notes in the left hand margin, to help orientate you and emphasise new and important messages.

BIOTOL will be used by universities, polytechnics and colleges as well as industrial training organisations and professional bodies. The texts will form a basis for flexible courses of all types leading to certificates, diplomas and degrees often through credit accumulation and transfer arrangements. In future there will be additional resources available including videos and computer based training programmes.

Preface

Based on predictions of population growth and the average income *per capita*, it is estimated that world food production will have to increase by at least 1.4 to 2.1 percent per annum. Central to responding to the need for increased food production, is to find ways of increasing crop yields per unit area of cultivated land. Many regard the emergence of plant biotechnology as being of crucial importance. The techniques of contemporary biotechnology enables the genetic capabilities of plants to be modified and provides routes to improvements in the maintenance of soil fertility and in the protection of crops from damaging environmental influences.

The setting of achievable biotechnological objectives to improve crop productivity can only be developed on the basis of a sound understanding of the physiological and biochemical characteristics of plants and on a knowledge of the effects a wide variety of physical, chemical and biological parameters have on crop productivity. This text provides a link between understanding the physiology of plants and the development of sound objectives for the application of biotechnology to the production of crops. It builds upon the assumption that readers are familiar with the physiological and biochemical properties of plants (covered in a partner BIOTOL text, 'Crop Physiology'), and examines the factors which influence crop yields. It begins by describing the measurements that are made in defining and evaluating crop productivity.

In Chapter 2, the internal factors which influence yield are discussed with special emphasis placed on the distinction between C_3 and C_4 plants.

In Chapters 3 and 4 we turn the attention of readers to the influence the external environment has on crop productivity. Thus, in Chapter 3 we examine the effect of physical and chemical factors (climate and soil fertility) whilst in Chapter 4 the biological factors (pests and diseases) are discussed.

Crops and plants are not, however, only cultivated in open-field. They may be produced in controlled environments within greenhouses and growth rooms or may be cultivated hydroponically. These systems of plant cultivation are described in Chapters 5 and 6.

Although the theme of the text centres on the production of food crops, plants are important for the production of a wide variety of commercially-valuable, non-food products. We especially think of medicines, flavourings and dyes in this context. Readers' knowledge of the diversity and production of these so-called secondary products will be enlarged by the material covered in Chapters 7 and 8.

This text does more than simply provide a practical context in which biotechnologists may set realistic objectives for improving crop production. By describing the core factors which influence crop productivity both in open-field and controlled environments, it provides the basis for developing sound practices for all engaged in the cultivation of plants.

Scientific and Course Advisors: Professor MCE van Dam-Mieras
Dr CK Leach

The measurement of growth, yield and productivity

The measurement of growth, yield and productivity

1.1 Introduction

crops The production of crops for consumption by people and livestock is central to the societies and economies of both developed and emerging countries. With world population forecast to double from its 1990 level of 5 billion by the early part of the next century, and with the growth expected to be largely in the developing countries, there is an urgent need to increase agricultural production from its present levels. An understanding of how yield (amount of crop produced per unit area) can be improved is essential in any scheme to increase total crop productivity.

Crop productivity is determined by many factors. These factors are either environmental processes acting on the plant population within the crop, or are components of the internal physiological processes of the crop species, particularly the fixation of carbon from atmospheric CO_2 into organic compounds by photosynthesis in the green tissues of the plant. These factors will determine the rate at which plants grow within the crop and hence determine the final yield of that crop. This chapter will review our knowledge of the components that make up yield in crop plants, and assess techniques to measure plant growth and crop productivity.

1.2 Yield and biomass

Some definitions

yield The term 'yield' is a general, non-specific term used by farmers and agronomists. They use it to signify a measure either of the total amount of a crop produced within a given area (eg within national or continental boundaries) or per unit land area within a local area (eg tonnes ha^{-1} of cereal grain). For detailed analysis of crop yields, more specific terms and parameters have been adopted:

- **biomass** of plants (W) - the total weight of plant material (above and below ground parts) per unit of ground area, at a given point in time;

- **productivity** (production) - the rate at which biomass (or organic matter) is accumulated by a crop, per unit of land area, per unit of time.

$\prod$ From a field trial plot of 1 m^2, you harvest 600 g of wheat. Express this yield in a) g m^2, b) kg ha^{-1} and c) tonnes ha^{-1}. What units would be most meaningful to describe your yield to a farmer?

Your answers should be:

a) 600 g m^{-2}, b) 6000 kg ha^{-1} and c) 6 tonnes ha^{-1}, since 1 ha = 10^{4}m^2 and 1 tonne = 1000 kg.

Although the SI units for mass, area and time are grams (g), m^2 and seconds (s), farmers (and agronomists) are more interested in the yield per growing area, so production would typically be expressed as kg ha^{-1} yr^{-1}, or tonnes ha^{-1} yr^{-1}.

1.2.1 The role of photosynthesis - more production parameters

biomass
productivity

The chemical energy stored in the organic compounds which has been fixed into the crop's tissues through photosynthesis is a major factor governing biomass productivity. Consequently, the level of crop productivity is determined by 4 main factors:

- the irradiance level ie the level of light energy available from sunlight;

- the amount of light energy intercepted by the crop stand;

- the efficiency of the plant in converting light to chemical energy;

- loss of chemical energy through plant respiration (and other losses eg grazing, disease).

production

This overall process of the conversion of light energy to chemical energy in the form of organic compounds is called primary production and can be divided into:

- **gross primary production** (Pg) - the total amount of organic matter fixed by a crop per unit ground area, per unit time (inclusive of respiratory losses);

- **net primary production** (Pn) - the total amount of organic matter assimilated as above, less that lost by respiration (R) (+ other losses), so that:

$$Pn = Pg - R.$$

∏ Can you think why biomass is different from production - after all they are both measures of yield!?

Biomass represents the quantity of biomass present at a single given time point, whereas production has a rate value representing the increase in the quantity of biomass over a given time period.

1.2.2 Effects of other environmental factors on production

photosynthesis

Although primary production is directly determined by the conversion of light energy into chemical energy (as stored in organic assimilates made during photosynthesis), there are a range of other environmental factors which, to varying degrees, can alter production.

∏ Make a list of the environmental factors you think would affect the rates of primary production.

effects of
latitude on
available solar
energy

Firstly, the amount of light (solar) energy can vary. Latitude is important, in this respect; tropical areas receive approximately twice the solar radiation that temperate areas receive. In temperate regions irradiance varies much more with the season, both in absolute level and in its duration (photoperiod), with both being reduced in the Winter months. Local meteorological conditions can also alter the amount of solar energy received, especially if the sun becomes obscured during overcast weather. Ultimately, the solar energy received over the course of the lifetime of a crop determines the total potential production of a crop (ie the maximum biomass possible at harvest).

factors which influence the ability of plants to harvest light energy

Other environmental factors can affect production by altering the rate at which the plants within the crop can harvest light energy. Such factors would include temperature, humidity, water availability, soil nutrient levels, pests and diseases. These particularly affect production through their effect on plant growth rates. Their effects are complex and will be discussed in greater detail in later chapters, but the importance of the relationship between individual plant growth rate and crop production will be stressed later in this chapter. The ability to quantitatively measure the effects on production of environmental factors has been central to the improvement of yields in modern agricultural systems.

1.2.3 Economic yield

Another way to report yield is to take account of the portion of the plant that is to be used, as in Table 1.1. Table 1.1 reports global yields of various groups in terms of the usuable parts of the crops.

Crop	Yield (x 10^6 tonnes)	Crop	Yield (x 10^6 tonnes)
cereals		*roots*	
wheat	510	potatoes	300
rice	465		
barley	178		
		others	
beverages		pulses	49
		grapes	63
coffee	6	sugar beet and	
tea	2.3	sugar cane	97
cocoa	1.8	cotton	17
		tobacco	6.6

Table 1.1 World yields in 1985 for a range of important crop species in million metric tonnes (data taken from the FAO production yearbook (1985)).

∏ Examine Table 1.1 and try to decide what part of the plant is being referred to in each case.

Some are easier to decide than others. The figures for cereals refer to grain weight. Potatoes and grapes are obvious, they refer to potatoes (ie the tubers) and grapes (ie the fruit). Pulses refer to the fruits of legume plants, such as peas and beans. Coffee and cocoa refer to dried beans, whereas tea and tobacco refer to leaves. Sugar beet and sugar cane refer to roots and stems respectively and cotton yield refers to the weight of cotton bolls. This is the name given to the cotton fruit which is covered with hairs that are used to make cotton fibres. These economically important parts of the plant are often referred to as the economic yield.

economic products and economic yield

It is evident that a range of economic products comes from crop species. The economic yield can be from either above ground or below ground plant parts, and the economic yield forms only a proportion of the total biological biomass at harvest. The economic yield may refer to the dry mass of the products, as in grain production, or to the fresh mass, as in lettuces and peas. The harvest component may need further processing to establish the true economic yield (eg oil from oilseed crops or sugar from sugar beet). In such cases, not only the total usable biomass, but the percentage of that biomass which represents the component of value, is important. Finally, biomass may be used for non-food purposes such as the production of fibre (eg cotton) or firewood or for industrial purposes.

Harvest Index As a measure of the proportion of a crop that is of economic use, the Harvest Index can be calculated. The relationship between Harvest Index, economic yield and total yield is given by:

$$\text{Harvest Index} \ = \ \frac{\text{economic yield}}{\text{total yield}} \ \text{x} \ 100\%$$

In most crops, which are non-root crops, total yield is normally considered to be the above ground parts of the crop because it is difficult to harvest the root material from most plants. The Harvest Index gives the proportion of the crop that is of economic use. It is of use in crop improvement strategies as an indicator of the efficiency with which assimilates are partitioned into the economically useful part of the crop.

1.3 Measurement of biomass and net production in field situations

1.3.1 Field trials

Measurements of biomass and net production are generally made in experiments performed in agricultural situations. The crop or crops of interest are grown in small plots which are intended to represent the performance of that particular crop species as if it was being grown on a much larger scale and, therefore, comparable to what might be found in the farmer's field. These plots tend to be of the order of a few square metres in area, and treatments (the application of an experimental variable) can be applied to each individual plot in accordance with the objectives of the experiment. For example, the effect of different amounts of nitrogen fertiliser on the yield of several wheat cultivars could be assessed in order to tell growers what the optimum level of application is. Several plots of each cultivar would be grown in the same field and to each plot a set amount of N-fertiliser would be added and biomass recorded. Comparison can also be made between experiments performed at widely separated locations to test the relative performance of cultivars in a range of climatic conditions.

field trials All such experiments are called field trials. A range of experimental designs has been developed for field trials to ensure that data from them are statistically valid.

∏ Before reading on, you might like to think of the precautions you would need to take in the design of a field trial such that you could be sure that the differences you find are due to the treatments you have given to your plots. Perhaps you could write down a few ideas. The following sections will help you to decide whether or not your ideas are sound.

Before considering experimental designs of field trials, some of the practical problems associated with experiments have to be addressed.

1.3.2 Randomisation

It is a good experimental practice to choose for the overall environment in which one's experiment is going to be performed, an environment that is as uniform as possible. This would mean that differences in the effects of treatments will not be masked by differences caused by the external environment being variable. Unfortunately, many fields selected as field trial sites show localised differences in one or more environmental factor, eg soil nutrients content, soil water content, proximity to hedgerows and so on.

For example, if one was testing the effect of phosphate fertiliser application in a field which has poor drainage at one end, and all the high phosphate treatments were conducted at the end with poor drainage, the effect of the high phosphate application might be masked by the excess water present. In consequence the objective of the experiments, *viz* to test the effect of the fertiliser, would be obscured. In practice, there may be no choice and this may be the only field available for the trial. To overcome the effects of differences in environmental conditions, field trial designs generally randomise plots throughout the selected site to remove the effects of these differences.

SAQ 1.1	If a particular field is known to show good and even drainage in all its parts, would there still be a need to randomise plots? Give a reason for your answer.

1.3.3 Replication

Within a field trial, not only are treatment positions randomised but it is also normal to have more than a single plot for each treatment, that is several plots are treated in the same way. Replication of plots allows more meaningful averages (means) to be calculated for each treatment, which provides a more reliable estimate of whether the applied treatment has had an effect on the response of biomass or production.

Field trials generally are costly to perform because they need quite large areas of ground and substantial labour input. Therefore, the amount of replication and the area of individual plots has to be reconciled to the cost of the field trial. Techniques are available to estimate what minimum amount of replication and plot area are needed to achieve the objectives of the experiment. A description of these is largely beyond the scope of this chapter but in the following section, we will consider some aspects of the design of a field trial.

1.3.4 Field trial design

As an example, let us consider testing 4 varieties (A to D) of sugar beet for biomass, replicating each variety 4 times. There are two types of commonly used field trial design:

Fully randomised design

Consider the following block representing four varieties (A-D) grown in a randomised arrangement of plots:

A	D	B	A
B	C	A	D
B	A	B	C
C	D	C	D

The four varieties are tested in a single field trial using 16 separate plots as shown above. The overall shape of the entire trial can be varied such as below and the area and shape of each individual plot can also be varied.

A	D	B	D	B	A	C	B	D	A	C	A	C	D	B	C

random
numbers

In every case, the position of each variety is determined randomly using either random numbers generated by computer or from a book of random number tables. One way to do this would be to assign numbers to the varieties eg 1-4 for variety A, 5-8 for variety B, 9-12 for variety C and 13-16 for variety D. The 16 plots would be numbered sequentially 1-16. Random numbers would then be selected using a two-digit code and ignoring all numbers above 16.

The first number selected, eg 7, would mean that a plot of variety B would be in position 1. If number 11 was selected next, a plot of variety C would be in position 2. This process would be continued until all plots of all varieties are assigned and it could be used for either of the trial shapes above. This technique requires that repeats of a number are discarded, as they would be if they were being chosen out of a hat.

This design is particularly suitable for fields where there is no obvious gradient in soil or environmental conditions.

Randomised block design

If a fully randomised design was used in a field with a wood on one side, it is possible that the effect below would be obtained.

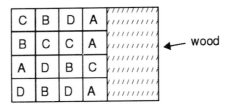

Thus most of the plots of variety A would be affected by the shade of the wood and it would be difficult to separate the effects of shade and A, even though randomisation had been used. This problem could be overcome by dividing the trial area into sub areas, called blocks, in such a way that the conditions within a particular block are as uniform as possible, as shown below.

```
                    block
               1   2   3   4
          1 |  A | D | C | D |/////////
          2 |  C | A | A | B |/////////  ← wood
    plot  3 |  B | C | B | A |/////////
          4 |  D | B | D | C |/////////
```

Each variety would appear in each block, its position being chosen at random, with a separate randomisation for each block. This design is called a randomised block design and it is the most frequently used design.

SAQ 1.2	How would you randomise the 4 varieties within a block?

Both fully randomised and randomised block designs allow the testing of the interaction between 2 factors. For example, the effect of the rate of fertiliser application on a range of crop varieties could be assessed. This is the topic of the next SAQ.

SAQ 1.3

You are asked to assess the yield effects of 3 levels (20, 60 and 150 kg ha^{-1}) of a newly formulated nitrogen fertiliser on the two wheat varieties, Norman and Hobbit. Preliminary trials suggest that 3 replicates of each treatment are needed in plots of 4 m^2. The only field available has a gradient in soil water content towards its eastern edge. Use this information to design a suitable field trial. We suggest you draw a plan of your field trial on a sheet of paper.

A common way to do the trial discussed in SAQ 1.3 would be to hire some spaces in a farmer's wheat field and employ the farmer to prepare the soil in the usual way for wheat. This ensures that the conditions for the trial are as realistic as possible.

1.4 Collection of data

Paths are often left between plots to allow field workers access to take measurements during crop development. In general, only plants growing in the central area of a plot are used for analysis because of edge and neighbour effects. Plants growing at the edge of a plot or block may grow faster because they have greater access to light. Therefore, they do not reflect what would be happening within the crop stand as a whole. Adjacent plots may contain plants of different height so that competition occurs along the edge of each plot, with similar effect.

selection of plants

Part of the experimental design involves deciding how to sample the experimental plants. We have already mentioned that we would avoid the outer rows of plants. Would we harvest all the others? The answer to this is sometimes yes, but frequently we harvest only a portion of them. The method used to choose which plants to harvest has to be decided very carefully. In field trials of wheat the plots are usually big enough to use a 1 m quadrat, arbitrarily placed to avoid edge plants, and to harvest all plants within the quadrat. Controlled environment trials usually contain fewer plants due to constraints on space and the experiment is usually designed so that all the plants are harvested. It is usually advisable to consult a statistician when deciding upon a design for the first time.

Measurements generally assess some component of biomass. Generally total biomass and economic yield at final harvest are the most interesting parameters, but measurements of biomass may be made at any time during the development of the crop. Harvested material is generally dried at 80°C in an oven, before dry weight is measured. For measurement of net productivity, organic content is often required. Organic content is determined by subtracting inorganic ash weight (obtained by burning dry samples in a muffle furnace at 600°C for 6 hours) from the dry weight.

statistical analysis

Field trial data is almost always subjected to statistical analysis, in order to properly evaluate the significance of the data. It is beyond the scope of this text to provide you with instruction in statistical analysis but you should be under no illusion as to its importance. For this reason, we have recommended several suitable texts at the end of this book in the 'Suggestions for Further Reading' section. If you are unfamiliar with statistical analysis, we strongly recommend you acquire the ability to carry out such procedures as they are of vital importance, not only to interpreting field data concerning crop yields but also in many other areas of science and technology.

As an example, we have used a statistical approach to evaluate data in an intext activity at the end of this chapter. There are many different approaches (for example regression analysis, t-tests, chi-tests etc) which have value in particular circumstances. Analysis of variance (ANOVA) tests are commonly used in the evaluation of field data.

Knowledge of statistical analysis is, however, not only important in interpreting data, it is also of great importance in the design of experiments. It helps us, for example, to decide how many samples to take and what size the samples should be.

1.5 Analysis of growth

plant growth

So far, we have looked at productivity only in terms of the production of biomass and final yield at the field level. In this section, we try to break down net productivity (the rate at which biomass accumulates) to allow a more detailed description of how plants grow within a crop stand. Although knowledge of the final yield in response to a treatment is of great use to the agronomist, of similar value is information about how a treatment affects plant growth (and hence final yield) during the development of the crop from germination to final harvest. Such a study is termed growth analysis.

Growth analysis studies can be divided into:

controlled environments

• individual plant studies on samples of single plants taken from field trials or grown in greenhouses or controlled environment cabinets;

field trial

• whole crop studies performed on whole or parts of plots in field trials.

We will deal with studies based on samples of single plants in Section 1.6 and then move on to studies based on crops in Section 1.7.

1.6 Techniques for plants treated individually

classical growth analysis

The most widely used way to analyse growth on single plants is called classical growth analysis . This type of analysis is now outlined with the following parameters:

• relative growth rate (RGR);

• unit leaf rate (ULR);

• leaf area ratio (LAR).

We will examine each in turn.

1.6.1 Relative growth rate

The most important parameter in classical growth analysis is the relative growth rate (RGR or R) which is the rate of increase in plant dry weight relative to the total dry weight of that plant at a single given time point (the instantaneous growth rate). This concept was first introduced by Blackman in 1919 who likened plant production to the compound interest law used to calculate bank interest.

In practice, RGR is not measured at a single time point but estimated as a mean RGR between 2 time points (see Figure 1.1).

RGR is defined by the following equation:

$$RGR = \frac{dW}{dt}\frac{1}{W} = \frac{d(\ln W)}{dt}$$

(Equation 1.1)

For calculation purposes, and for our convenience as experimenters, this equation can be simplified to the following:

$$RGR = \frac{\ln W_2 - \ln W_1}{(t_2 - t_1)}$$

(Equation 1.2)

where W_1 and W_2 are plant weights at first and second harvests respectively. t_1 and t_2 are the times of the first and second harvests.

We can represent the RGR from both these equations graphically to demonstrate that the first equation (1.1) gives a value at one single time point- the instantaneous rate - whereas Equation 1.2 gives a mean value between two time points (Figure 1.1).

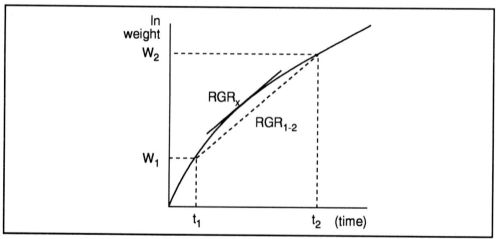

Figure 1.1 Measurement of mean and instantaneous growth rate. RGR_{1-2} is the mean relative growth rate between two time points (given by dotted line), RGR_x is the relative growth rate at any given time along the actual growth curve of the plant (solid line).

The graph above shows the derivation of mean RGR, and RGR at a single point, from a plot of ln W against time.

Experimentally, data for plant weight are easy to obtain by destructive harvesting of plants at the required time intervals (ie to measure W_1 and W_2 at t_1 and t_2).

absolute growth rate

It is important not to confuse RGR with absolute growth rate (AGR), which is the total increase in weight over a time period (eg g week^{-1}). Thus:

$$AGR = \frac{(W_2 - W_1)}{(t_2 - t_1)}$$

SAQ 1.4	From the data below for 2 different plants, calculate the RGR and AGR for each plant and record your answer in the boxes. Which plant is growing the fastest over this time period?

Quantity	Plant 1	Plant 2	Units
W_1	2	20	g
W_2	5	30	g
t_1	14	14	days
t_2	21	21	days
$W_2 - W_1$	☐	☐	g
$\ln W_2 - \ln W_1$	☐	☐	
$t_2 - t_1$	☐	☐	days
AGR	☐	☐	☐
RGR	☐	☐	☐

RGR refers to changes in biomass (or organic content) during the life cycle of the crop. This gives us a great deal of information about plant performance for comparison between varieties or treatment, but it does not give information about the performance of the assimilatory system, which drives the accumulation of organic matter (chemical energy) and biomass. The assimilatory system is, of course, mainly located in the leaves which are the principal photosynthetic organs.

Therefore in growth analysis, two types of measurement are usually made:

- plant weight (dry weight or organic content);

- size of the assimilatory system (leaf area or sometimes leaf protein or chlorophyll).

This allows the calculation of 2 further parameters which assess the assimilatory capabilities of the plant.

1.6.2 Unit leaf rate

ULF or
NAR

The unit leaf rate (ULR or E) is defined as the increase in plant biomass per unit of assimilatory area (leaf) and is often called Net Assimilatory Rate (NAR).

As for RGR, an equation is available to calculate the instantaneous ULR for a given single time point in a plant's life cycle.

$$ULR = \frac{dW}{dt}\frac{1}{A}$$

(Equation 1.3)

where W = plant weight, t = time and A = leaf area.

Note that this is the unit leaf rate at one particular time (the instantaneous unit leaf rate).

In practice of course, the leaf area, A, will change with time and the unit leaf rate is calculated as a mean value between two time points. This may be calculated from Equation 1.4:

$$ULR = \frac{(W_2 - W_1)(\ln A_2 - \ln A_1)}{(A_2 - A_1)(t_2 - t_1)}$$
(Equation 1.4)

where A_1 and A_2 are total leaf area at harvests 1 and 2.

ULR has typical units of $g\ cm^{-2}\ day^{-1}$ or $kg\ m^{-2}\ week^{-1}$.

1.6.3 Leaf area ratio

Leaf area ratio (LAR or F) is the ratio between the assimilatory area (leaf area) and total biomass at any given time point.

It is given by the equation:

$$LAR = \frac{A}{W}$$
(Equation 1.5)

where A = leaf area and W = plant weight.

LAR is an instantaneous measurement, but is generally required with RGR and ULR which are normally calculated between two time points. The mean LAR can be calculated between 2 time points by the following equation:

$$\text{rate increase in LAR} = \frac{(A_2 - A_1)(\ln W_2 - \ln W_1)}{(W_2 - W_1)(\ln A_2 - \ln A_1)}$$
(Equation 1.6)

SAQ 1.5

A plant variety has been grown and samples harvested 35 and 42 days after sowing. The mean plant weights and leaf areas were determined. This analysis gave the following results:

Day 35 harvest   W = 1.0 g, A = 106 cm^2;

Day 42 harvest   W = 3.9 g, A = 467 cm^2.

Use this data to calculate:

1) the mean ULR value for the period 35-42 days for the plant variety;

2) the LAR values of the plant on days 35 and 42;

3) the average (mean) value of LAR in the period 35-42 days.

Note on units: units are generally selected for calculated growth parameters as appropriate for the experimental regime. For example, if small plants are harvested at weekly intervals, units based on tonnes per year would be inappropriate whereas g per day or week would be more meaningful.

1.6.4 Relationship between RGR, ULR and LAR

You may notice if you look carefully that the basic equations for these three parameters share common terms. In some cases, a relationship holds between them as below:

$$RGR = ULR \times LAR \qquad \text{(Equation 1.7)}$$

Π You might like to prove the relationship shown in Equation 1.7 for yourself. Begin by thinking about the increase in plant dry weight using our relative growth rate and the increase in leaf area using our unit leaf rate.

Here is our solution:

since:

$$RGR = \frac{\ln W_2 - \ln W_1}{(t_2 - t_1)} \qquad \text{(see Equation 1.2)}$$

therefore:

$$t_2 - t_1 = \frac{\ln W_2 - \ln W_1}{RGR}$$

and:

$$ULR = \frac{(W_2 - W_1)\,(\ln A_2 - \ln A_1)}{(A_2 - A_1)\,(t_2 - t_1)} \qquad \text{(see Equation 1.4)}$$

Therefore:

$$t_2 - t_1 = \frac{(W_2 - W_1)(\ln A_2 - \ln A_1)}{(A_2 - A_1)ULR}$$

then:

$$\frac{\ln W_2 - \ln W_1}{RGR} = \frac{(W_2 - W_1)\,(\ln A_2 - \ln A_1)}{(A_2 - A_1)ULR}$$

thus:

$$\frac{RGR}{ULR} = \frac{(\ln W_2 - \ln W_1)\,(A_2 - A_1)}{(W_2 - W_1)\,(\ln A_2 - \ln A_1)}$$

Equation 1.6 shows that:

$$LAR = \frac{(A_2 - A_1)(\ln W_2 - \ln W_1)}{(W_2 - W_1)(\ln A_2 - \ln A_1)}$$

therefore:

$$\frac{RGR}{ULR} = LAR, \text{ or } RGR = ULR \times LAR$$

Equation 1.7 appears to be more reliable for instantaneous data, but not so for data calculated for two harvests.

∏   If the units used for RGR are g g^{-1} day^{-1} and the units for ULR are g cm^{-2} day^{-1}, what are the appropriate units for LAR?

The answer is cm^2 g^{-1} since LAR is the ratio of leaf area (cm^2) per unit of plant mass.

You can check that this is correct using:

$$\frac{RGR}{ULR} = LAR$$

Substituting in units $\dfrac{g\,g^{-1}\,day^{-1}}{g\,cm^{-2}\,day^{-1}} = cm^2\,g^{-1}$.

∏   What would be the units of LAR if the units being used for RGR kg kg^{-1} day^{-1} and the units for ULR are kg m^{-2} day^{-1}?

The answer is m^2 kg^{-1}.

1.6.5 Experiments with RGR, ULR and LAR

sequential harvest

In practice, experiments on sequential harvests of groups of individual plants are made to assess RGR, ULR and LAR. Each plant from each treatment is measured separately for dry weight and leaf area. A problem with this approach is that calculations then have to be performed on the meaned results for each treatment harvest. This does not give any information as to the statistical variability of RGR, ULR or LAR.

A full statistical treatment of this is beyond the scope of this chapter. However, here we indicate in outline, the sort of procedure that may be followed. The process often used is to pair plants from two harvests of the same treatment (eg smallest plant harvest 1 with smallest plant harvest 2, next smallest plant harvest 1 with next smallest plant harvest 2, etc) and calculate the differences in dry weight and leaf area for each harvested pair. From this, an estimate of the statistical variability can be made (standard deviation, standard error or analysis of variance - ANOVA).

1.6.6 Alternatives to classical growth analysis

RGR, ULR and LAR are the three products of classical growth analysis and are calculated from data taken at fixed time points. Thus, this approach is an 'interval' approach. ULR and LAR have both been criticised because the relationship between biomass and leaf area differs between species through differences in leaf thickness.

Classical growth analysis has been extended so that the persistence of leaf biomass and leaf area is incorporated. Obviously, as a plant develops, the length of time that a leaf has assimilatory power is important. For this, indices of biomass duration and leaf area duration have been developed. Despite criticisms, classical growth analysis remains the most widely used and useful form of growth analysis.

functional approach

A second approach, called the functional approach, compares computer generated growth curves which have been fitted to the raw growth data. The functions obtained can be compared to demonstrate differences between treatments.

A third approach, developed by Harper and his co-workers for studies of natural vegetation, is to compare treatments on a demographic basis, using information about population numbers, rather than using biomass and leaf area, as the basis of comparison.

Both these approaches are specialised areas and beyond the scope of this text.

1.7 Techniques for plants treated as crops

crops grown in stands

The previous descriptions have revolved around plants handled as individuals, but the principles of classical growth analysis are easily modified to analyse growth in stands of plants (eg field plots). It is particularly the responses of crops growing in the field which are of interest in terms of yield and production.

The main difference between experiments on individual plants and crops is that field trials are performed on a specified area. This has led to the modification of classical growth analysis to allow growth analysis on an area basis (area generally being given the term, P).

Leaf Area Index and Crop Growth Rate

The 2 parameters devised for crop growth analysis are the Leaf Area Index (LAI), a ratio between the area of leaves in the crop and the ground area occupied, and the Crop Growth Rate (CGR). The Crop Growth Rate is a measure of the increase in crop biomass per unit time. Usually it is measured in terms of increase in biomass per unit time per unit area. The CGR is, therefore, akin to an absolute growth rate. Compare this with the Relative Growth Rate (RGR) we defined earlier. RGR is the increase in crop biomass per unit of biomass already present per unit time.

1.7.1 Leaf Area Index

crop leafiness

LAI gives an estimate of the leafiness of a crop within a field trial plot and represents the area of leaf for a given land area.

$$LAI = \frac{A}{P}$$

(Equation 1.8)

where A = leaf area and P = plot area.

Often it is inconvenient and time consuming to measure the total leaf area of a plot, so the leaf area of several representative plants is taken and the total plot leaf area is then calculated from the density of the plants in the plot. Units for LAI are typically m^2 (leaf) m^{-2} (ground area), and represent a value at a single time point in crop development. You will, of course, note that LAI is an instantaneous measurement.

1.7.2 Crop growth rate

In previous sections we indicated that the instantaneous RGR is given by:

$$RGR = ULR \times LAR$$

(see Section 1.6.4). Similarly, we can write:

$$CGR = ULR \times LAI.$$

LAI is an instantaneous value and replaces LAR in the calculation. CGR calculated in this fashion is an instantaneous value with units of weight per unit land area, per unit time.

In practice, CGR is conveniently measured from plot harvests separated by a time interval using the relationship:

$$CGR = \frac{W_2 - W_1}{t_2 - t_1}$$

(Equation 1.9)

where W_1 and W_2 are the harvested biomass from field plot. Of course, they may come from different areas, so W values have to be corrected to the same area before use in this equation.

In other words, for W_1 and W_2 we should use values of biomass per unit area.

SAQ 1.6

From field plots of different sizes, barley was harvested 40 and 60 days from sowing. The biomass at 40 days was harvested from 1 m^2, had a leaf area of 1.3 m^2 and a biomass of 75 g. At 60 days, from an area of 1.5 m^2, the leaf area was measured at 3.3 m^2 and biomass was 200 g. Calculate the CGR and the instantaneous LAI for each harvest, giving appropriate units.

1.8 Yield components

A different, but equally useful method of evaluating biomass (particularly final yield) of a crop, is to construct a model of how reproductive, developmental and morphological features of plants in crop stands contribute to final yield.

This is perhaps best illustrated with a couple of examples.

We will use cereal crops in our first example. Remember that each cereal plant within a stand consists of a number of shoots called tillers, which may or may not carry an ear of corn.

Thus we could report yield in terms of the amount of grain by using:

		number of plants per unit area	x	mean number of tillers with ears per plant	x	mean number of grains per ear	x	mean grain weight
yield/unit area	=							

This approach is useful because it allows us to evaluate what effect each component has on yield. In this way, the agronomist or plant breeder is able to optimise each factor to the maximum benefit of the farmer.

A more complex example has been developed for pod beans:

		number of plants per unit land area	x	mean leaf area per plant	x	mean plant biomass per unit leaf area	x	mean pod biomass per unit plant biomass
yield	=							

In common with the cereal example, each factor can be varied in a field trial in order to optimise each component to give maximum yield.

1.9 Case study: assessment of yield improvements in UK wheat varieties released between 1900 and 1980

Look at the information in Table 1.2. It comes from an actual field trial carried out in the UK in 1977 to test whether improvements in the yield of wheat in the UK during this century were due to genetic improvement in the varieties on offer to the farmer. It is of considerable importance to a plant breeder to be able to establish that he is actually breeding a new variety that has greater yield potential than the previous varieties on offer.

To test this, a series of varieties released to the farmer over the course of this century were provided, and the varieties were grown in a randomised block design, with 4 blocks and each variety replicated once in each block. The individual plot size was 4.2 m x 1.4 m.

Mean values (ie the mean of 4 replicates) for a series of yield and growth parameters are given for the varieties grown.

Variety	Year of Intro-duction	Yield t/ha	Harvest Index %	Mean grain weight (g)	No. grains /ear	No. ears /m^2	LAI	Date of flowering June
Little Joss	1908	5.2	36	38.8	36.8	366	6.9	20
Holdfast	1935	5.0	36	32.6	32.5	468	6.6	19
Capelle-Desprez	1953	5.9	42	44.8	30.1	435	7.4	18
Maris Widgeon	1964	5.7	39	45.1	30.3	415	8.1	17
Maris Huntsman	1972	6.5	46	47.7	36.2	379	8.2	16
Hobbit	1977	7.3	48	36.9	46.1	429	6.3	12
Mardler	1978	6.2	49	38.2	42.1	386	7.6	12
Norman	1980	7.6	51	47.5	43.6	366	5.9	13

Table 1.2 Data relating to wheat varieties released in the UK between 1900 and 1980.

∏ Examine these data and see if you can draw some conclusions to the following questions.

Do newer varieties have higher yields than older varieties? (Yes/No). If yes, see if you can identify parameters that have contributed to the improved yield in new varieties.

It might prove helpful to plot graphs of yield, Harvest Index, etc against year of introduction as these will provide a more visual illustration of the data. (Do this for yourself before reading on).

Examination of the data presented indicates that the yield (in tonnes ha^{-1}) are higher in the newer varieties than in the older ones. The data relating to Harvest Index also show a progressive increase in the newer varieties.

The identification of the parameters which have contributed to improved yields is less easy.

Let us consider each in turn.

Mean grain weight and parameters relating to the number of grains

We have plotted mean grain weight and yield against year of introduction in Figure 1.2a. In Figure 1.2b, we have plotted yield against mean grain weight. It is difficult to discern a positive relationship between increased yield and mean grain weight from the data plotted in this figure. This is supported by the use of statistical evaluation. In this case we have used a regression analysis but, there are many other tests that could be used.

In Panel 1.1 we briefly remind you of this type of analysis.

To determine whether or not there is a relationship between the sets of values (x, y) we first calculate a correlation coefficient (r) from the relationship:

$$r = \frac{\Sigma(x - \bar{x})(y - \bar{y})}{\sqrt{\Sigma(x - \bar{x})^2 \Sigma(y - \bar{y})^2}}$$

where $\bar{x}$ and $\bar{y}$ are the means of the two sets of variables.

If there are n sets of variables, then the system is said to have n - 1 degrees of freedom. We then use appropriate statistical tables relating r values, degrees of freedom and the probability that a linear relationship exists between x and y.

In our case we have sets of paired data corresponding to the 8 varieties used. Thus we have 7 degrees of freedom. With 7 degrees, r values need to be 0.75 for us to be 95% certain that a linear relationship holds between two sets of variables. To be 99% certain that a linear relationship holds, r must be 0.798.

Panel 1.1 A brief reminder of the determination of correlation coefficients to determine if there is a linear relationship between two variables.

Returning to our intext activity, an r value of 0.799 was determined for yield against year of introduction, indicating a good correlation between these two. The more recently introduced varieties giving better yields. On the other hand, the correlation coefficient (r) for yield and mean grain weight is only 0.436, indicating that there is little evidence for a direct relationship between these two variables.

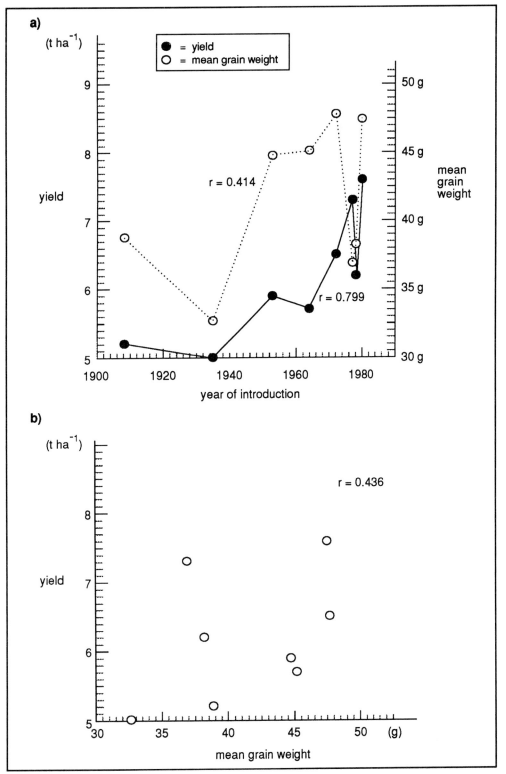

Figure 1.2 a) Plots of yield and mean grain weight against year of introduction. b) Plot of yield against mean grain size (see text for discussion). The correlation coefficient (r) has been calculated for each relationship.

There appears to be some evidence of a relationship between yield and number of grains per ear (Figure 1.3), but not for the number of ears per unit ground area (Figure 1.4). These data do not provide convincing evidence that these parameters have had a positive effect on yield. Note we have reported r values on the Figures.

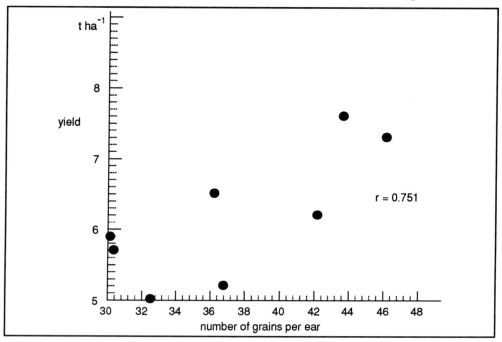

Figure 1.3 Plot of yield against number of grains per ear (see text).

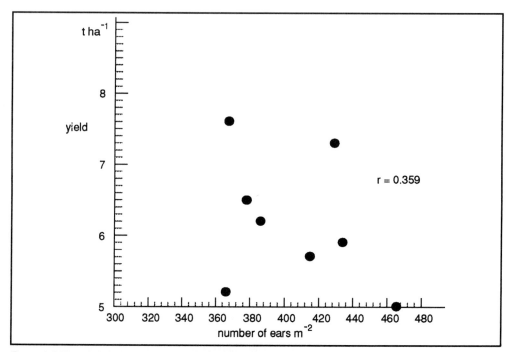

Figure 1.4 Plot of yield against number of ears (see text).

The relationship between LAI and yield, shown in Figure 1.5, also shows little evidence of a positive correlation between these two parameters.

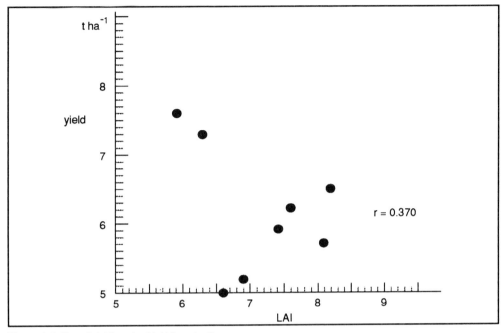

Figure 1.5 Plot of yield against LAI (see text).

Date of flowering and yield

Plots relating yield and date of flowering show a significant correlation coefficient and indicate that the earlier flowering varieties tend to have higher yields (Figure 1.6). Table 1.2 shows that newer varieties tend to be earlier flowering. It is possible that the earlier flowering varieties have a longer period in which to assimilate organic matter to fill their grain. This is borne out by the correlation between date of flowering and the Harvest Index illustrated in Figure 1.7.

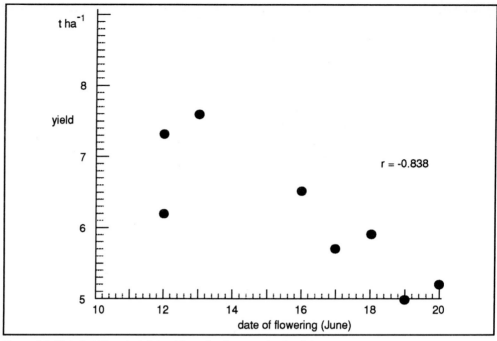

Figure 1.6 Plot of yield against date of flowering (see text for details).

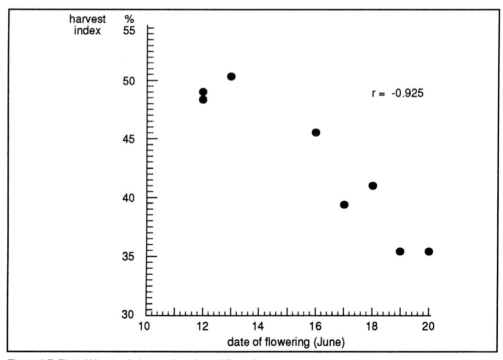

Figure 1.7 Plot of Harvest Index against date of flowering.

This type of analysis would indicate that it may be more beneficial for plant breeders to concentrate their efforts towards developing earlier flowering varieties of wheat than attempting to generate varieties with increased grain numbers, increased grain size or increased number of ears.

Summary and objectives

In this chapter we have outlined concepts relating to crop yield and differentiated between biomass and crop production. We have briefly outlined the range of environmental factors which influence production. The main emphasis in this chapter has been on the design of experiments and the measurements that are made to evaluate yield and other crop characteristics.

In the final part of the chapter, we showed how the detailed measurement of the various components of yield may help to identify breeding targets and lead to further increases in yield.

Now that you have completed this chapter, you should be able to:

- define yield, biomass and production;

- explain what is meant by growth and by assimilation;

- be able to design appropriate experimental and measurement strategies to evaluate growth and yield;

- be able to use supplied data to calculate rates of growth, ULR, LAR and LAI;

- identify relationships between yield and other measured parameters from supplied data.

Internal factors affecting the primary productivity of plants

Internal factors affecting the primary productivity of plants

2.1 Introduction

This text has been written on the assumption that the reader is familiar with the main physiological properties and mechanisms that operate within plants. (These are described in the partner BIOTOL text 'Crop Physiology'). Nevertheless, photosynthesis and respiration play such a key role in the crop productivity that these issues are worth revisiting and extending here.

In this chapter, we briefly examine the processes of CO_2 assimilation including the mechanisms of light harvesting. We also contrast the assimilation of CO_2 in the two main photosynthetic groups, the so called C_3 and C_4 plants. We will then consider respiration processes in crop plants and consider the consequences of these in terms of their effects on crop production. This discussion will provide a framework in which to examine the effects of environmental factors on crop productivity dealt with in subsequent chapters.

2.2 CO₂ assimilation processes

It is convenient to examine the processes of CO_2 in three stages. These are:

- diffusion processes, in which CO_2 moves from the atmosphere towards the site of assimilation;

- the photochemical processes in which the energy of light is entrapped and converted to utilisable biochemical energy and reducing power;

- the biochemical processes, in which CO_2 is reduced using the energy and reducing power generated in the photochemical processes.

2.2.1 CO₂ diffusion to the site of assimilation

The pathway of diffusion of CO_2 from the air is through the stomata into the air spaces in the leaf (and in some cases, the stem) tissues. From there, it diffuses through the cell walls and protoplasm of the mesophyll cells to the chloroplasts. We have stylised this pathway in Figure 2.1.

∏   Examine Figure 2.1 and see if you can identify what factors may influence the rate of diffusion of CO_2 to the chloroplasts.

The main factor is the difference in the concentration of CO_2 in the air and the various compartments of the leaf. There are several boundaries (barriers) that the CO_2 must pass in order to reach the chloroplasts. We have represented these in Figure 2.2.

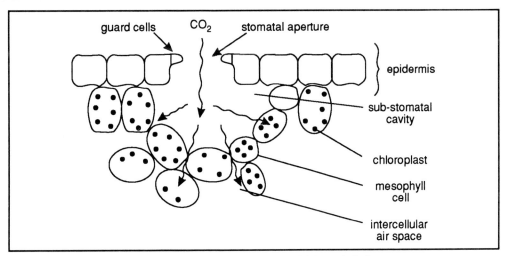

Figure 2.1 Pathway of diffusion of CO_2 from the air to the site of CO_2 assimilation.

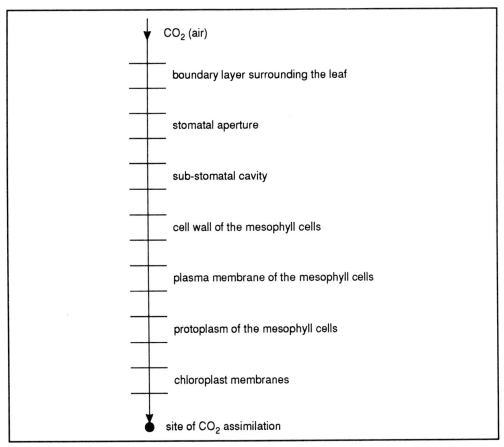

Figure 2.2 Barriers which CO_2 must diffuse through to reach the site of CO_2 assimilation.

Let us examine these barriers in a little more detail. The air on the surface (boundary) layer of the leaf may, because of CO_2 uptake by the plant tissues, have a lower CO_2 concentration than that of the atmosphere in general. Therefore CO_2 will diffuse into the layer. Thus the CO_2 concentration at the stomatal aperture may be lower than that of the atmosphere in general and will be governed by the rate of diffusion into this region. We can regard this boundary layer as forming a resistance to CO_2 transfer to the chloroplasts. Similarly, there is restricted access to the air spaces within the leaf. The rate of transfer of CO_2 into these air spaces will be influenced by the number and size of the stomatal apertures. We can, therefore, regard the presence/absence of stomata as offering a further resistance to CO_2 transfer to the chloroplasts. Analogous resistances to CO_2 transfer operate at the other boundaries illustrated in Figure 2.2

Therefore if we plot CO_2 concentrations along the pathway of CO_2 transport, we can anticipate a picture of the type illustrated in Figure 2.3.

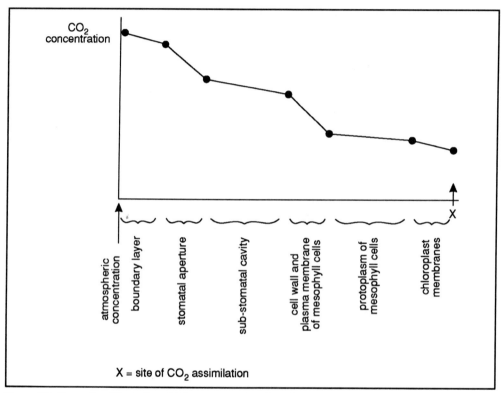

Figure 2.3 Generalised illustration of CO_2 concentration along the pathway of CO_2 transfer from the atmosphere to the site of CO_2 assimilation.

Of course, the exact shape of this concentration profile will be dependent upon many factors.

Π   See how many such factors you can list.

The sort of factors you may have included are:

• Wind speed. The boundary layer represents a thin layer of non-turbulent (laminar) air flow. The thickness of this layer (and hence its resistance to CO_2 flow) is dependent upon wind speed.

- Presence of hairs on the leaf. These will increase the thickness of the boundary layer and increase the resistance.

stomatal resistance

- Stomatal resistance. This depends upon the number of stomata and the size of their apertures. The apertures of stomata are under continual change and are greatly influenced by the availability of water. Water shortage causes stomata to close to reduce water loss through transpiration. Under such conditions, stomatal resistance to CO_2 transfer is high and thus CO_2 supply to the site of assimilation is reduced. This, in turn, results in low assimilation rates.

mesophyll resistance

- Composition and organisation of mesophyll cells. We can see from Figure 2.3 that a series of resistances operate between the air space in the plant tissues (stomatal cavities) and the site of CO_2 assimilation. These resistances will depend on the physical arrangement of the cells and on the chemical composition and organisation of the mesophyll cell walls, plasma membranes, protoplasm and chloroplast membrane. We may regard these as being predominantly determined by the genetic properties of the cells. They are, however, also influenced by environmental factors. For example, loss of turgor of the mesophyll cells as a result of water shortage will alter mesophyll cell properties which, in turn, will influence the resistance to CO_2 transfer. It is, perhaps, best to group the serially-linked resistances to transfer between air spaces and chloroplast as a single resistance, the mesophyll resistance.

Thus, the diffusion of CO_2 to the site of CO_2 assimilation is driven by concentration differences and is restricted by a variety of resistances. The extent of these resistances depends upon both the properties of the plant and upon environmental factors.

We can regard the flux of CO_2 into the site of CO_2 assimilation as being related to the concentration differences and the resistance to CO_2 flux. Thus:

$$CO_2 \text{ flux is proportional to } \frac{\text{concentration difference}}{\text{resistance}}$$

In a mathematical form, we can write this as:

$$\phi_{CO_2} = \frac{[CO_2]_{atm} - [CO_2]_{ch}}{r_b + r_s + r_m}$$

where:

ϕ_{CO_2} = net rate of CO_2 diffusion into the leaf [g (CO_2) m^{-2}, (leaf) s^{-1}]

$[CO_2]_{atm}$ = CO_2 concentration in the atmosphere [g (CO_2) m^{-3}]

$[CO_2]_{ch}$ = CO_2 concentration in the chloroplasts [g (CO_2) m^{-3}]

r_b = resistance to CO_2 diffusion in the boundary layer around the leaf (s m^{-1} = time per unit distance)

r_s = stomatal resistance to CO_2 diffusion (s m^{-1})

r_m = mesophyll resistance to CO_2 diffusion (s m^{-1})

Typical values for these resistances are:

$r_b \sim 20 \text{ s m}^{-1}$

$$r_s \sim 50 - 10\,000\,\mathrm{s\,m^{-1}}$$

$$r_m \sim 100 - 1000\,\mathrm{s\,m^{-1}}$$

SAQ 2.1

1) From the three resistances to CO_2 flux described in the text, which resistance is most variable?

2) Is a value of $10\,000\,\mathrm{s\,m^{-1}}$ for stomatal resistance most likely to be encountered under wet conditions or under conditions of water limitation?

3) When stoma are fully open, which resistance is likely to limit CO_2 flux most?

2.2.2 Photochemical processes

The absorption spectra of the major plant pigments are shown in Figure 2.4.

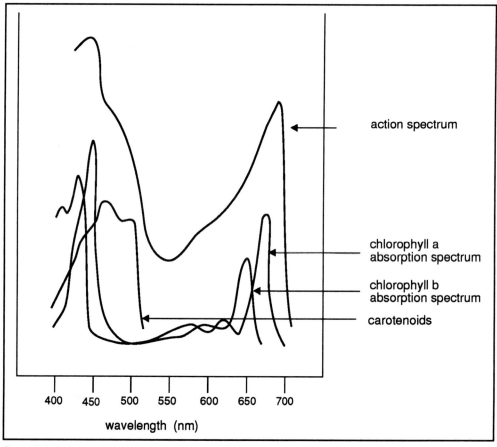

Figure 2.4 Action spectrum of photosynthesis and absorption spectra of the major photosynthetic pigments.

Also reported in this figure is the action spectrum of photosynthesis, ie the rate of photosynthesis that occurs under different wavelengths of light, for a typical plant. The key point to realise is that plants only use radiation with wavelengths between 400 and 700 nm in the photochemical process. Light of these wavelengths is often referred to as photosynthetically active radiation (PAR) and constitutes roughly half of the radiation

photosynthe-
tically active
radiation (PAR)

reaching the Earth's surface. We will be referring to PAR in more detail in Chapter 3. For now, think of it as radiation that reaches the Earth's surface that can be absorbed by the photosynthetic pigments of plants.

We will not examine the biophysical processes of photosynthesis in detail here. They are best dealt with in the context of plant physiology (see for example the BIOTOL text 'Crop Physiology'). However, the stoichiometry of the overall photochemical reaction (known as the light reaction) is well established and can be written as:

$$\text{light quanta} + 2H_2O + 2NADP^+ + 3ADP + 3Pi \rightarrow$$

$$2NADPH + 2H^+ + 3ATP + 3H_2O + O_2$$

(Note that 3 molecules of water are released by the formation of 3 molecules of ATP from ADP and Pi).

In other words, the energy of light is used to drive the production of ATP + NADPH. Although some differences are observed between plants, there is a considerable degree of uniformity concerning the efficiency of the primary energy harvesting process.

The utilisable source of energy (ATP) and reducing potential (NADPH) are, however, used with differing efficiencies depending on whether the C_3 or the C_4 route of CO_2 assimilation takes place. We will examine each in turn.

2.2.3 C_3 route of CO_2 assimilation

Table 2.1 gives examples of so called C_3 crop species. You will note that these include both monocotyledonous and dicotyledonous plants.

barley	broad beans
peas	potato
rice	sugar beet
wheat	

Table 2.1 Examples of C_3 crop species.

CO_2 assimilation by these plants is dependent totally upon the ATP and NADPH generated by the photochemical process.

In C_3 plants, the enzyme RuBisCo (ribulose bisphosphate carboxylase) catalyses the primary CO_2 assimilation reaction. Ribulose 1-5 bisphosphate acts as the primary CO_2 acceptor and the primary product is phosphoglycerate. It is from this 3 carbon product that the C_3 plants derive their designation.

The products of CO_2 assimilation are then rearranged via the Calvin cycle, to regenerate the acceptor molecule, ribulose 1,5 bisphosphate. A simplified version of this process is shown in Figure 2.5.

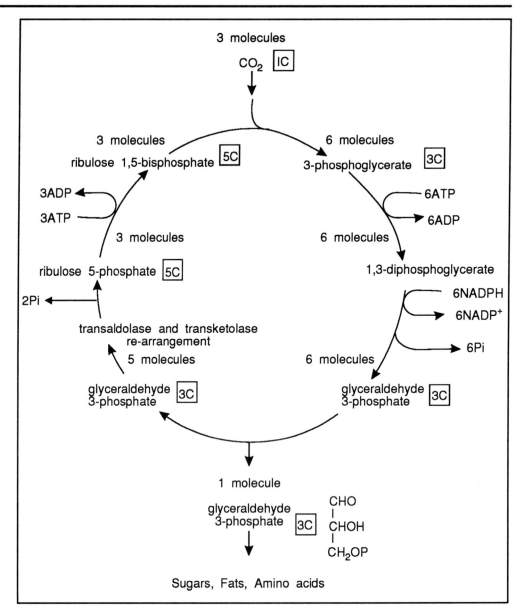

Figure 2.5 The Calvin cycle.

Π Use the information shown in Figure 2.5 to write an overall reaction for the process. Start with 3 molecules of CO_2. Remember that chemicals that are formed in one reaction but consumed in another, will not appear in the overall reaction.

You should have been able to write the overall reaction as:

$$3CO_2 + 9ATP + 6NADPH \rightarrow \text{glyceraldehyde 3-phosphate} + 8Pi + 6NADP^+ + 9ADP$$

photo-
respiration

You should also note that RuBisCo can react with O_2. The enzyme may become oxygenated and inactive. Energy is consumed in restoring the enzyme. Also, in the presence of oxygen, RuBisCo catalyses a reaction leading to the production of phosphoglycolate. This is metabolised by a pathway leading to the release of CO_2. This process is referred to as photorespiration and represents both a loss of energy harvested during the photochemical events and a loss of assimilated CO_2. (The biochemistry of this process is described in the BIOTOL text, 'Crop Physiology', so we will not elaborate on it further here). It should be self-evident that photorespiration reduces the overall efficiency of the photosynthetic assimilation of CO_2. The extent of this reduction is dependent upon the conditions, especially on CO_2 and O_2 concentrations. Under typical atmospheric conditions (0.035% v/v CO_2 ; 21% v/v O_2), photorespiration may reduce CO_2 assimilation by 30% or more. In O_2 rich or CO_2 poor conditions (for example after stomatal closure caused by excess transpiration), photorespiration may have an even stronger effect.

We must conclude, therefore, that the efficiency of C_3 photosynthesis in terms of CO_2 assimilation is greatly influenced by environmental conditions.

2.2.4 C_4 route of CO_2 assimilation

Table 2.2 gives some examples of C_4 crops.

maize	sorghum
millet	sugar cane
sisal	

Table 2.2 Examples of C_4 crops.

C_4 plants

These plants are called C_4 plants because the first stable product of carbon reduction is a 4-carbon molecule (oxaloacetate). C_4 plants produce two types of chloroplasts. These are the large chloroplasts of the parenchymatous cells which surround the vascular bundles (known as bundle sheaths) and the small chloroplasts of the mesophyll cells that occupy the remainder of the leaf space.

role of
mesophyll cells
and bundle
sheath cells in
C_4 plants

In the chloroplasts of the mesophyll cells, phosphoenol pyruvate (PEP) carboxylase catalyses the primary step in CO_2 assimilation. The product, oxaloacetate, is converted to malate and transported to the cells of the bundle sheaths. Here the malate is converted to pyruvate and CO_2. The CO_2 is re-assimilated but this time via the Calvin cycle within the large chloroplasts of the bundle sheaths. The pyruvate formed in the cells of the bundle sheaths diffuses back into the mesophyll cells where it is re-converted to phosphoenol pyruvate at the expense of ATP. These processes are illustrated in Figure 2.6.

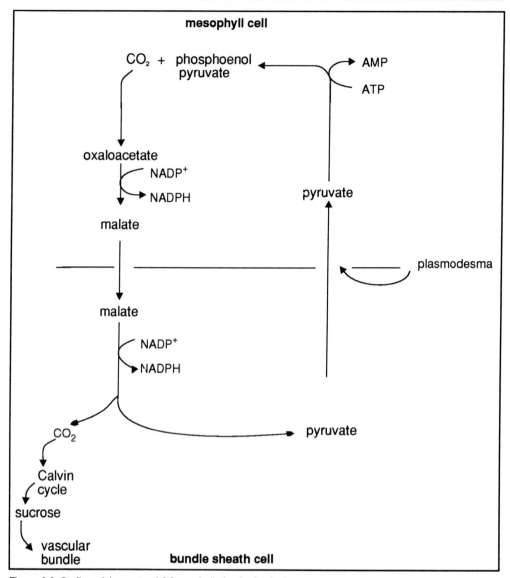

Figure 2.6 Outline of the route of CO_2 assimilation in C_4 plants.

Π On the basis of the information given so far, is photosynthesis in C_3 plants more
   or less efficient than that of C_4 plants?

Based on the information you have been given, you would conclude that C_3 plants are
most efficient. In both C_3 and C_4 plants, the ultimate CO_2 assimilation mechanism is via
the Calvin cycle. This presumably operates at a similar efficiency in both plants.
However in C_4 plants there is an energy consuming process for delivering CO_2 to the
site of the Calvin cycle. For each molecule of CO_2 delivered, one molecule of ATP is
converted to AMP (see Figure 2.6). This ATP is supplied by dark respiration.

In practice however, if we measure CO_2 assimilation efficiency in terms of CO_2 fixed m^{-2}
of leaf surface per unit time, we find that C_4 plants are more efficient than C_3 plants.
How can we explain this apparent paradox? In outline, it can be explained by the
observations that the phosphoenol pyruvate (PEP) carboxylase of the C_4 mesophyll cells

has a higher affinity for CO_2 than the RuBisCo of the chloroplasts of the C_3 mesophyll cells and that PEP carboxylase does not react with O_2. Furthermore, the CO_2 shuttled to the bundle sheath cells as malate, results in high concentrations of CO_2 in the bundle sheath cells. This favours reaction of the RuBisCo with CO_2 rather than O_2 and reduces photorespiratory losses. We will examine the consequences of this on assimilation rates in Section 2.3.

2.3 Influence of environment on carbon dioxide assimilation

In Chapter 3, we will be considering the effects of environmental factors on biomass production by whole plants and communities of plants. Here, we are predominantly focusing on the effects of environmental parameters on individual plant tissues (mainly leaves). We will consider light irradiance, CO_2 concentration and temperature.

2.3.1 Effects of light irradiance on CO_2 assimilation

There have been many experiments in which CO_2 assimilation, has been measured as a function of light. A stylised example is shown in Figure 2.7. In this figure we have shown the net assimilation of CO_2 by a C_3 leaf in normal air. You will note that, at high values, further increases in light irradiance are not matched by corresponding increases in CO_2 assimilation rates. In other words, the system is saturated with light. With lower light intensities, decreases in light intensity are however, matched by decreases in CO_2 assimilation rates. You will note that the net CO_2 assimilation-light irradiance response curve does not go through the origin. There is a low light intensity at which there is no net CO_2 assimilation. This is the so-called light compensation point. This does not mean that no CO_2 assimilation is taking place, merely that the rate of CO_2 assimilation is balanced by the rate of CO_2 released by respiratory processes.

light compensation point

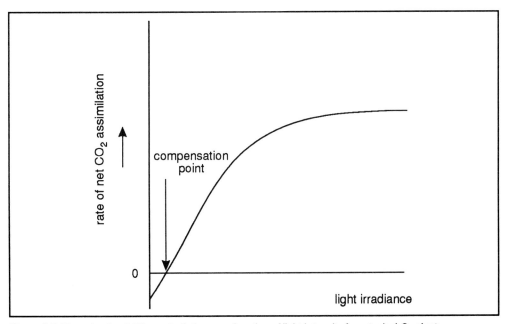

Figure 2.7 The rate of net CO_2 assimilation as a function of light intensity for a typical C_3 plant.

Let us take this analysis a little further. Consider a leaf of a C_3 plant in total darkness. Such a leaf would be carrying out respiration to generate a usable form of energy to drive transport and other essential processes (see Section 2.4). Under these conditions, there would be a net release of CO_2 (that is, a net dissimilation of CO_2). If a light was switched on, the dissimilation process would continue but would be counteracted by the photosynthetic assimilation of CO_2. However, as the light irradiance increases, so does the rate of photorespiration. Thus, we can regard the net rate of CO_2 assimilation as being the difference between rate of gross CO_2 assimilation rate and the rate of gross CO_2 dissimilation. This latter rate is made up of two components, dark respiration and photorespiration. Dark respiration is independent of light intensity whilst photorespiration is light dependent.

We can, therefore, write for net CO_2 assimilation rates that:

net CO_2 assimilation = gross CO_2 assimilation - CO_2 dissimilation

or

net CO_2 assimilation = gross CO_2 assimilation - (dark respiration + photorespiration)

Symbolically we can write this as:

$$A_{net} = A_{gross} - (R_{dark} + R_{photo})$$

where A = assimilation rates, R = respiration rates.

SAQ 2.2

In Figure 2.8, we have plotted net CO_2 assimilation, dark respiration and photorespiration rates at different light irradiances.

Add to this figure a plot of the relationship between CO_2 assimilation and light irradiance, for a leaf incubated in an O_2-free atmosphere.

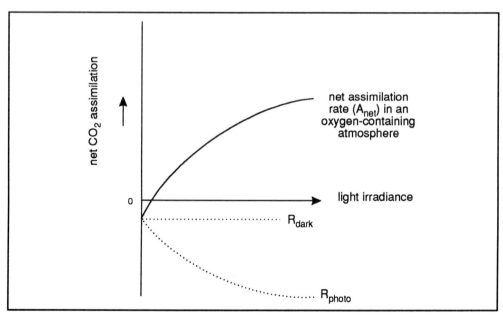

Figure 2.8 Plots of net CO_2 assimilation, dark respiration and photorespiration rates at different light irradiances (see SAQ 2.2).

Photorespiration is a negative contributor to plant production. However, as we pointed out earlier, this is predominantly a feature of C_3 plants.

Therefore, if we plot graphs of CO_2 assimilation against light irradiance for C_3 and for C_4 leaves, we might anticipate the sorts of relationships shown in Figure 2.9.

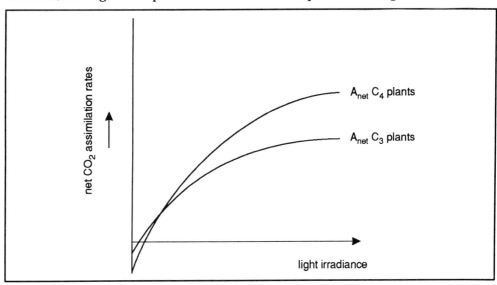

Figure 2.9 Net assimilation rates for C_3 and C_4 leaves as a function of light irradiance.

It must be remembered that the principal difference between the two is the absence of photorespiration in C_4 plants.

SAQ 2.3   Figure 2.9 indicates a lower net CO_2 assimilation in C_4 leaves at low light irradiances than in C_3 leaves. Explain why this is so.

If we plot gross assimilation rates against light irradiances, then curves of the type shown in Figure 2.10 are produced.

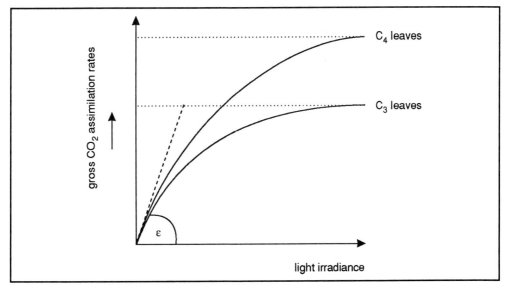

Figure 2.10 Stylised representation of gross CO_2 assimilation for C_3 and C_4 leaves as a function of light irradiance. Note that the initial slope is referred to as ε.

Note that at low light irradiance, the gross assimilation rates for C_3 and C_4 leaves are very similar. At these low light irradiances, the availability of light energy is rate limiting. The slope of the curves at these very low light values (ε; see Figure 2.10) gives a measure of the initial light use efficiency. ε values are very similar for both C_3 and C_4 leaves indicating that process inefficiencies (for example photorespiration in C_3 leaves and energy used for CO_2 transport in C_4 leaves) are very similar.

light saturation

If, however, the light irradiance is increased then the process of photorespiration in C_3 plants reduces the efficiency of these leaves relative to that of C_4 plants. As light irradiances are increased even further, process inefficiencies (for example photorespiration and the diffusion of CO_2 to the sites of photosynthesis) become more pronounced. Eventually a situation is reached in which CO_2 assimilation rates reach a plateau (A_{max}). We can describe leaves under these conditions as being light saturated. The values of A_{max} vary between species. Table 2.3 reports some typical values.

	A_{max} {kg CO_2 ha(leaf)$^{-1}$ h^{-1}}	ε {kg CO_2 ha(leaf)$^{-1}$ h^{-1}/J m^{-2} s^{-1}}
C3	40	0.45
C4	70	0.45

Table 2.3 Typical values of A_{max} and ε for C_3 and C_4 species.

∏ Note the units used in Table 2.3 for both A_{max} and ε. See if you can suggest some alternative units.

There are many different units you could have suggested. A_{max} should be expressed in terms of: amount of CO_2 assimilated/area of leaf/unit time. Thus g m^{-2} s^{-1} would be suitable. ε is given by:

$$\frac{\text{assimilated } CO_2 \,/\, \text{area of leaf} \,/\, \text{unit time}}{\text{irradiance}}$$

Thus it should have a similar unit as we use for A_{max}, divided by units of radiation intensity where units of radiation intensity = energy/unit area/unit time. Most instruments give a direct read out as J m^{-2} s^{-1} but different areas and times could be used.

The reason we have dwelt on the use of units for these values is that the choice of units depends on the scale of process under consideration. Clearly when we are considering processes involving a single leaf used in an experiment, the unit area under consideration is of the order of cm^2 and the time scale may be minutes or hours. On the other hand, if we wish to consider CO_2 assimilation in an agricultural setting, units of hectares and days may be more appropriate.

∏ Do all the leaves on a plant display the same values of A_{max} and ε? Justify your answer.

not all leaves on a plant are the same

The answer is no. Leaves adapt to the light conditions under which they develop. In a typical plant, some leaves (the lower ones) may be predominantly in shade while the upper leaves will be predominantly in light. Thus leaves may, for example, be thicker or have a higher RuBisCo content. They may also contain more, or less, pigments other than chlorophyll (for example carotenoids) which may not transfer the energy they absorb from light to the photochemical production of ATP and NADPH. Thus we must anticipate that all leaves of a particular metabolic class (that is C_3 or C_4) will not show the same values of A_{max} or ε. Generally, leaves produced under high light intensities show higher maximum CO_2 assimilation rates than those produced in low light.

In many natural situations, light use efficiency may be reduced because of increased photorespiratory losses and because diffusion of CO_2 becomes the limiting factor. We will examine this in the next section.

2.3.2 Influence of CO_2 concentration on CO_2 assimilation rates

In the previous section, we indicated that, at high light intensities, CO_2 assimilation may be limited by CO_2 supply. We might anticipate, therefore, that incubation of leaves in atmospheres enriched with CO_2 would lead to higher rates of CO_2 assimilation. This is indeed what happens (see Figure 2.11). We might further anticipate that by doubling the concentration of CO_2 in the atmosphere then, since the rate of diffusion is proportional to the concentration gradient, the rate of CO_2 assimilation would be doubled. In many situations such a simple proportionality between A_{max} and external CO_2 concentration is not observed.

∏ See if you can reason why this might be so (the equation relating the CO_2 flux to diffusion given in Section 2.2.1 might help you).

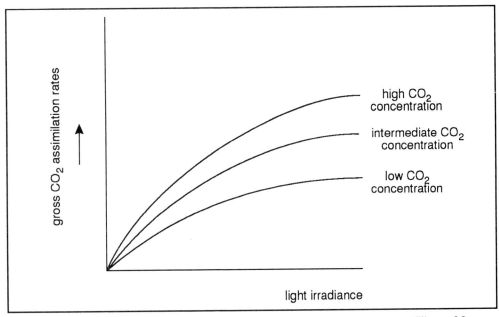

Figure 2.11 Stylised relationships between CO_2 assimilation rates and light irradiance at different CO_2 concentrations. Typical CO_2 concentrations would be in the range 300-2500 ppm (note 1 ppm = 1 μl CO_2 l^{-1} air = 1.83 mg CO_2 m^{-3} air at typical ambient temperatures).

The rate of CO_2 diffusion to the site of photosynthesis is influenced by the concentration gradient(s) of CO_2 across the various boundaries (see Figure 2.2) and by the resistance of the boundaries. We described this in Section 2.2.1 by the relationship:

$$\phi_{CO_2} = \frac{[CO_2]_{atm} - [CO_2]_{ch}}{r_b + r_s + r_m}$$

Assuming that the resistances are constant, ϕ_{CO_2} should be proportional to $[CO_2]_{atm} - [CO_2]_{ch}$.

The failure to get direct proportionality between ϕ_{CO_2} and external CO_2 concentration ($[CO_2]_{atm}$) could suggest that one (or more) of the resistances is itself influenced by external CO_2 concentrations, and that is indeed the case.

In many instances, the stomatal aperture (and thus stomatal resistance, r_s) is influenced by external CO_2 concentration. High external CO_2 concentration usually causes the stomata to close thereby increasing r_s. Thus attempts to continually increase A_{max} by simply increasing external CO_2 concentration may be, in the end, counter productive as the high CO_2 concentration may, indirectly, limit the supply of CO_2. There is, however, a substantial difference between C_3 and C_4 plants in the effects of increasing CO_2 concentration on CO_2 assimilation rates as is illustrated in Figure 2.12.

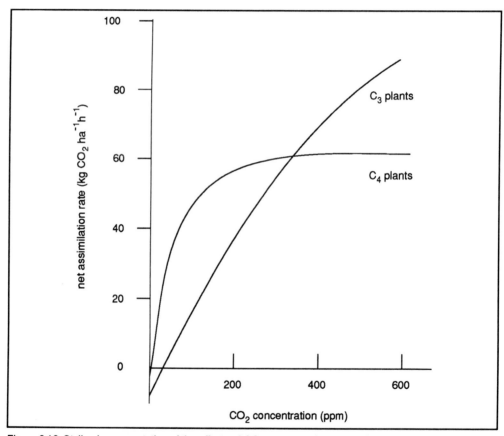

Figure 2.12 Stylised representation of the effects of CO_2 concentration on net CO_2 assimilation rates in C_3 and C_4 plants. These data were obtained under light saturating conditions and CO_2 concentration was measured in the sub-stomatal cavity.

∏ Why is the CO_2 concentration in the sub-stomatal cavity, rather than in the atmosphere, measured in experiments of the type illustrated in Figure 2.12?

By using the CO_2 concentration of the sub-stomatal cavity we obtain a more direct relationship between CO_2 concentration and CO_2 assimilation. If we had used the atmospheric concentration of CO_2, then we would be uncertain about the CO_2 concentration in the vicinity of photosynthesis. Recall, also, that external CO_2 concentration influences stomatal responses. By measuring the CO_2 concentration in the sub-stomatal cavity, we avoid confusing stomatal responses with changes in CO_2 concentration.

∏   Re-examine Figure 2.12 and see if you can explain why C_3 and C_4 plants show different responses to CO_2 concentration.

The data presented in Figure 2.12 show that assimilation rates in C_3 plants increase more gradually than those in C_4 plants. The primary CO_2 acceptor system in C_3 plants is RuBisCo, whilst in C_4 plants it is phosphoenol pyruvate (PEP) carboxylase. RuBisCo has a lower affinity for CO_2 than has PEP carboxylase, therefore higher CO_2 concentrations are required to saturate RuBisCo. Furthermore, C_3 and C_4 plants have quite different 'CO_2 compensation points'. The 'CO_2 compensation point' is the consentration of CO_2 when CO_2 assimilation just compensates for (balances) CO_2 dissimilation.

CO_2 compensation point

For C_4 plants this is typically of the order of 5-10 ppm CO_2, whilst for C_3 plants it is 5-10 times greater. Although a range of values have been determined experimentally, the mean value for C_3 plants has been determined as 120 ppm. Higher conpensation points are caused by the additional CO_2 released by photorespiration in C_3 plants. At higher CO_2 concentration, photorespiration is reduced and this leads to higher CO_2 assimilation rates. Thus, C_3 plants continue to show improved assimilation rates even at very high CO_2 concentrations. Ultimately however, a maximum will be reached or, alternatively, the very high concentrations of CO_2 may be toxic to some other cellular process(es).

The profile shown in Figure 2.12 for C_4 plants is typical of a system which has a high affinity for CO_2. Assimilation rates reach a maximum at a relatively low CO_2 concentration.

SAQ 2.4

The CO_2 in the atmosphere in a greenhouse is maintained at a level that establishes CO_2 concentrations in the sub-stomatal cavities of plants at about 200-300 ppm. It is proposed to increase the level of CO_2 in the atmosphere in the greenhouse. Will C_3 or C_4 crops, or both, benefit from such an action?

So far we have established that both light irradiance and CO_2 concentration have profound effects on rates of assimilation. Now we will turn our attention to the influence of temperature.

2.3.3 Influence of temperature on CO_2 assimilation rates

Since CO_2 assimilation is predominantly a biochemical process involving enzymes, we might anticipate that the CO_2 assimilation against temperature profiles would resemble those observed with enzymes. This is, in fact, what is observed.

In Figure 2.13, we have plotted composite activity-temperature profiles for RuBisCo from C_3 plants and PEP carboxylase from C_4 plants.

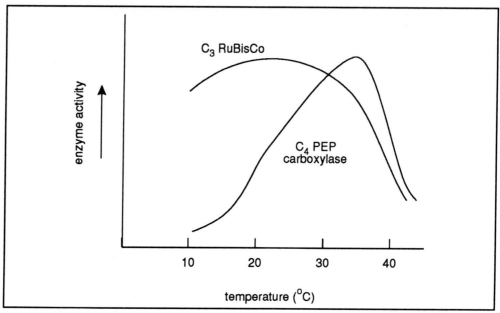

Figure 2.13 The effects of temperature on the *in vitro* activity of RuBisCo and PEP carboxylase.

∏ From the data presented in Figure 2.13, which group of plants (C_3 or C_4) are best adapted to grow in warmer regions?

You should have concluded that C_4 plants are best adapted to grow at higher temperatures.

∏ Is this supported by the information in Tables 2.1 and 2.2?

The answer is yes. The C_4 crops listed in Table 2.2 are predominantly those from tropical and sub-tropical regions. The C_3 crops listed in Table 2.1 include many examples from more temperate regions.

This generalised split between C_3 and C_4 plants is supported by the rates of CO_2 assimilation by leaves of plants at different temperatures (Figure 2.14).

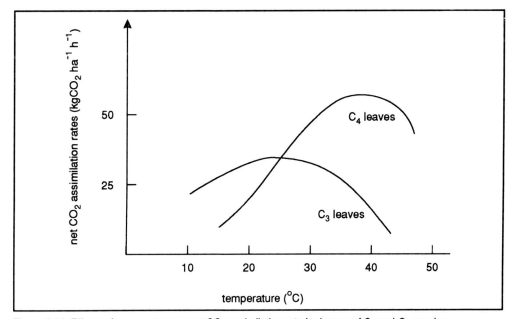

Figure 2.14 Effects of temperature on net CO_2 assimilation rate by leaves of C_3 and C_4 species.

This is as far as we need to take the discussion concerning CO_2 assimilation at this stage. You should, however, realise that other factors may greatly influence the net rate of CO_2 assimilation. For example, the availability of water may, through its effect on stomatal aperture (and, therefore, on stomatal resistance) greatly influence CO_2 concentration in the sub-stomatal cavities. Similarly, the nutrient status of the plant will also influence CO_2 assimilation rates. Deficiency of any essential nutrient limits the plant's ability to make the machinery of photosynthesis or use it effectively. Leaf age is also important. Very young leaves are not photosynthetically fully active, and leaves close to senescence have reduced photosynthetic capabilities. Pathogens and pests will also influence the extent of net CO_2 assimilation. We will discuss these factors further in the context of whole crops in Chapters 3 and 4. Here we will continue to focus on the internal capabilities of plants by considering the processes of respiration.

water and nutrient status influences photosynthetic efficiency

2.4 Respiration in plants

So far we have introduced you to two terms: photorespiration and dark respiration. In this section we will examine respiratory processes in plants, not from a biochemical perspective, but from the perspective of agricultural production.

growth and maintenance energy requirements

By respiration, in this context, we mean the oxidation of assimilated carbon materials with, or without, the production of a utilisable form of energy. From our perspective, energy is used for maintenance of the plant and for growth. We can, therefore, distinguish maintenance respiration and growth respiration. We will deal with each in turn.

2.4.1 Maintenance respiration

∏ Bearing in mind that photorespiration is the consumption of photosynthetic assimilates for the repair of oxygenated RuBisCo in C_3 plants, should this be regarded as a maintenance respiration?

maintenance
energy needed
for repair and
for the
maintenance
of
concentration
gradients
If we define the maintenance requirement as the energy that is needed to maintain the plant and to support its functioning, then we can, indeed, regard photorespiration as a form of maintenance respiration. In a wider context, part of the maintenance energy requirement is concerned with the replacement or repair of the chemical constituents of the plant. Chemical components of plants are typical of components of all living organisms: they have a limited lifespan. Their repair or replacement costs energy. The consumption of energy for these purposes, although essential, does not lead to any net increase in plant material. The opposite is, in fact, true. Also the consumption of energy to maintain concentration gradients (for example the uptake of nutrients, maintaining osmotic pressure and pH) is essential but does not lead to any direct increase in biomass. The spontaneous chemical hydrolysis of ATP (and other intermediates) not linked to metabolically useful reactions also represents an energy drain on the system for which there is no net biomass gain.

We use the term maintenance energy to encompass all of these types of activities. In other words:

$$
\begin{array}{c}
\text{maintenance} \\
\text{energy} \\
\text{requirement}
\end{array}
=
\begin{array}{c}
\text{energy} \\
\text{consumed to} \\
\text{replace/repair} \\
\text{constituents}
\end{array}
+
\begin{array}{c}
\text{energy consumed} \\
\text{to maintain} \\
\text{concentration} \\
\text{gradients}
\end{array}
+
\begin{array}{c}
\text{energy lost} \\
\text{through} \\
\text{spontaneous} \\
\text{chemical hydrolysis}
\end{array}
$$

SAQ 2.5

Would the maintenance energy requirement be increased or decreased if the plant was cultivated in the presence of:

1) a compound which increased the permeability of membranes to ions;

2) a pathogen such as a virus?

It is important in agricultural terms to minimise maintenance energy requirements but it is, however, difficult to quantify. The maintenance energy requirements are, of course, related to the amount of biomass. It is also dependent upon the composition and activities of the biomass. The maintenance energy will be different for different organs. It is, therefore, usual to use a pragmatic approach and to estimate the maintenance respiration rate (R_M) on the basis of the dry weight (W) of the various organs and the maintenance coefficient for each organ. The maintenance coefficient (α) is the rate of maintenance respiration per unit mass of the organ. Typical units are kg CH_2O per kg dry weight per day.

If we divide a plant into leaves, stem, root and storage organ, we can write the following empirical formula:

$$R_M = \alpha_{leaf} W_{leaf} + \alpha_{stem} W_{stem} + \alpha_{root} W_{root} + \alpha_{storage} W_{storage}$$

Typical values for the maintenance coefficients are:

$$\alpha_{leaf} = 0.03 \text{ kg } CH_2O \text{ kg}^{-1} \text{ dry weight day}^{-1}$$

$$\alpha_{stem} = 0.015 \text{ kg } CH_2O \text{ kg}^{-1} \text{ dry weight day}^{-1}$$

$$\alpha_{root} = 0.01 \text{ kg } CH_2O \text{ kg}^{-1} \text{ dry weight day}^{-1}$$

$$\alpha_{storage} = 0.01 \text{ kg } CH_2O \text{ kg}^{-1} \text{ dry weight day}^{-1}$$

SAQ 2.6	Explain why the maintenance coefficient for leaves is higher than those for stems, roots and storage organs.

It is useful to compare values for the maintenance coefficient for whole plants by taking into account the proportion of the plant made up by the various organs and the α values of these organs. Some typical maintenance coefficients for whole plants are reported in Table 2.4.

Crop	α (kgCH$_2$O kg^{-1}day^{-1})
oil rich seed crop	0.030
protein rich seed crop	0.025
cereals	0.015
root/tuber crop	0.010

Table 2.4 Maintenance coefficients from various crop types. The maintenance coefficients (α) reported here are for whole plants. Note that the values are for plants maintained at 20°C. Data from Deulen H van and Wolf J (1986), Modelling of Agricultural Production: Weather, Soils and Crops, Simulation Monographs, Wageningen.

Maintenance coefficients are dependent upon temperature, they usually double for every 10 degree rise, although there is some variability between crops.

SAQ 2.7	1) Calculate how much carbohydrate is consumed per day in maintenance respiration if you have a standing crop of 1000 kg potatoes. Assume that the mean ambient temperature is 20°C.
	2) What would be the daily consumption of carbohydrates (or their equivalent) in maintenance respiration of a 1000 kg crop of oil seed rape if the temperature is maintained at about 30°C?
	3) Roughly, what proportion of the biomass is consumed per day to provide maintenance energy in the situations described in 1) and 2)?

SAQ 2.8	Suggest a reason why the rate of net assimilation in Figure 2.8 would be expected to begin to fall over a period of time.

2.4.2 Growth respiration

Growth respiration is the process of respiration that provides the energy required for biomass synthesis. Part of the products of CO_2 assimilation are metabolised to provide the energy for the biosynthesis and construction of cellular materials used in growth. The amount of assimilates used for this purpose depends upon the composition of the end products. It is sometimes expressed as a growth cost usually in g of assimilate (glucose) used per g product. The product could be a general product eg biomass or a specific product such as fat, protein, or organic acids. An alternative to the 'growth cost' is the 'conversion efficiency' ie g product per g of assimilate used. The conversion

growth cost and conversion efficiency

efficiency is, in fact, the reciprocal of the growth cost. Its value is independent of temperature.

The growth cost for making particular chemical products includes a number of components. Firstly there are the energy costs of the individual reactions leading to the synthesis of the product. These can be calculated accurately if the metabolic pathways for their synthesis are known. The energy costs also include the energy costs of transporting precursors to the site of synthesis and products to the sites of storage (or use). Some examples of these for major groups of bioproducts are given in Table 2.5.

Compound	Biosynthetic costs (g glucose g^{-1} product)	Transport costs (g glucose g^{-1} product)	Growth cost * (g glucose g^{-1} product)
Carbohydrate	1.211	0.064	1.275
Proteins	1.824	0.096	1.920
Fats	3.030	0.159	3.189
Organic acids	0.0906	0.048	0.954

Table 2.5 Biosynthetic, transport and growth costs for major groups of bioproducts. *Growth costs are calculated by summing biosynthetic and transport costs. Data from Penning de Vries, et al (1989), Simulation of Ecophysiological Process of Growth in Several Annual Crops, Simulation Monographs, Wageningen.

∏ Calculate the conversion efficiencies of the compounds listed in Table 2.5.

The values are:

carbohydrates = 0.78 g prod g^{-1} gluc;

proteins = 0.52 g prod g^{-1} gluc;

fats = 0.31 g prod g^{-1} glu;

organic acids = 1.05 g prod g^{-1} glu.

These are calculated simply by taking the reciprocals of the growth costs. Note that for compounds that are less oxidised than glucose, the conversion efficiency is much lower than 1 (see fats). If the compound is more oxidised than glucose then the conversion efficiency may be greater than 1 (see organic acids). In this case, we are effectively adding oxygen to glucose so one gram of glucose may give rise to more than one gram of organic acid, despite the fact that some glucose will be completely oxidised to supply energy for this process.

CVFs   In many cases, these conversion efficiencies are referred to as conversion factors (CVF).

Conversion factors can also be determined for plant organs based on the chemical composition of the organs. Surprisingly, when this is done for a large number of plants, the mean values of the CVFs of leaves, stems and roots are very similar and are about 0.68g organ g^{-1} glucose. There is, however, a greater diversity amongst storage organs, reflecting their greater diversity in chemical composition. Some values are reported in Table 2.6.

Crop	Carbohydrate	Protein	Fats	CVF
Peanut	14	27	39	0.40
Soybean	29	37	18	0.46
Broad bean	55	29	1	0.57
Maize	75	8	4	0.67
Wheat	76	12	2	0.70
Potato	78	9	0	0.78

Table 2.6 Chemical composition (% dry weight) and conversion factors (CVF) of a variety of storage organs. Data from Penning de Vries *et al* (1989), Simulation of Ecophysiological Process of Growth in Several Annual Crops, Simulation Monographs, Wageningen.

∏ From the data presented in Table 2.6, see if you can formulate some general rules relating the chemical composition of storage organs to CVF values.

The general rules are that storage organs with high carbohydrate content and low protein and fat content have high CVFs (see for example potato, wheat and maize). Increases in the proportion of protein will tend to reduce the overall CVF (see for example broad bean). The storage of fats has, however, the greatest impact on CVF values (see for example peanut and soybean).

SAQ 2.9

Rank the following crop groups in ascending order of their anticipated CVF values:

cereals, protein rich seed crops, root/tuber crops, oil rich seed crops.

SAQ 2.10

Match up the statements in list A with those in list B.

A

1) Growth cost.

2) Maintenance coefficient

3) Conversion factor.

4) Growth respiration.

B

1) Repiration used for biomass biosynthesis.

2) Mass of assimilate used (g) per g of biomass (or product) produced.

3) Mass of biomass (or product) produced per g of assimilate used.

4) Rate of maintenance respiration per unit mass of the organ.

Summary and objectives

In this chapter, we have examined the main physiological factors which influence the productivity of plants. These physiological factors cannot, however, be understood without consideration of some of the environmental factors which influence them. We began by examining the CO_2 assimilation process including the diffusion of CO_2 to the site of photosynthesis, the photochemical processes of photosynthesis and the biochemical routes of CO_2 assimilation. We distinguished between two major groups of plants, the C_3 and C_4 plants, and explained the significance of photorespiration in C_3 plants. We described the effects of light intensity, CO_2 concentration and temperature on CO_2 assimilation. Towards the end of the chapter, we described the processes of respiration in plants in relation to their effects on yield. We distinguished between respiration linked to maintenance requirements and respiration linked to growth. We introduced the terms maintenance coefficients, growth cost, conversion efficiency and conversion factor.

Now that you have completed this chapter, you should be able to:

- describe the route by which CO_2 reaches the site of photosynthesis and explain the various resistances to this movement;

- describe the effects of light irradiance and O_2 concentration on gross and net CO_2 assimilation rates in C_3 and C_4 plants and explain the differences;

- explain how CO_2 concentration influences CO_2 assimilation rates in C_3 and C_4 plants;

- explain why plants need to generate maintenance energy and identify plant organs that are likely to have high maintenance requirements;

- use, appropriately, terms that apply to plant energy consumption, including maintenance coefficients, growth cost, conversion efficiency and conversion factor;

- calculate energy consumption and conversion factors from supplied data.

External factors affecting crop productivity: the physical environment

External factors affecting crop productivity: the physical environment

3.1 Introduction

At present, there is considerable pressure on farmers, especially in developing countries, to maximise crop yields. Of particular importance in affecting the yield of any crop are the environmental conditions the crop encounters during its growth and development. Environmental factors determine the rate at which a crop grows and influence whether a particular crop can be grown at a particular location. They are a major factor in governing final yield. In order for farmers to maximise yields, identification of those environmental factors that critically affect yield is necessary, together with a knowledge of how they modulate the growth and development of a crop.

In developing countries, with high population growth rates, the general improvement of yields is of concern. In Europe and North America, on the other hand, there is as much concern about reducing agricultural inputs (themselves 'environmental' factors), as agricultural pollution increases due to the over-application of fertilisers and pesticides. In each case, an understanding of how the environment factors interact with crop growth and yield can help us achieve increased yields and reductions in agricultural pollution.

3.1.1 Internal versus external factors

genetic
potential

At this point, it is important to identify the main components that affect the yield of a crop. Firstly, the inherent or internal properties of the crop itself will be important. This may be termed the genetic potential of a crop variety, and it will determine the physiological and biochemical potential of the variety. Under perfect growing conditions, the maximum yield of a crop variety is determined genetically because the genetic make up determines how efficient the overall physiological and biochemical processes of the plants are in producing useful biomass.

In field situations, perfect growing conditions rarely exist, and realistically it is the way a crop variety performs in response to a range of external factors, which we normally call environmental factors, that determines useful biomass and yield. In such situations, plant breeders and crop physiologists often work together to try to develop new combinations of genetic material suited to a given range of environmental factors, with the objective of maximising yield.

3.2 Environmental factors - the broad scale

∏ Make a list of all the environmental factors that you can think of that may affect the yield of: 1) a Middle-Eastern wheat crop and 2) a European apple crop.

We will not attempt to give you a complete answer to this at this stage but we hope that you would have considered such factors as: availability of nutrients in the soil, availability of water, day length, sunlight intensity (influenced by latitude), the extent of cloud cover, temperature (including fluctuations), winds (may cause physical damage to the plant) and so on.

the effects of
pests on yield

You may have included, also, the occurrence of plant pests and pathogens. For example, although the physical and chemical parameters of the environment might be suitable to achieve excellent yields of apple crops in Europe, the occurrence of large numbers of pests such as codling moths (*Laspreyresia pomonella*) may render the crop virtually unusable. Similarly large outbreaks of attack by members of the fungal genus *Claviceps*, which causes ergot infections in cereals, can destroy an otherwise high-yielding wheat crop.

The point we are trying to make is that there are a great many factors that affect crop growth and yield. In this chapter, it will be possible to give only an overview of all these factors, drawing your attention to the most important ones. We will give greater detail for selected situations in some case studies at the end of the chapter. Nevertheless, it is hoped that by the end of the chapter you will have grasped the fundamental principles of how environmental factors influence crop yields. This is a long chapter so it would be advisable not to attempt to study it all in one sitting.

3.2.1 Definition of an environmental factor

For our purposes, we can define an environmental factor as any property (physical, chemical or biological) of the surroundings of a plant or crop, which may interact with the growth, development and, for farmers, the harvestable yield of that plant or crop.

∏ For a wheat crop that, for a given stage of its development, has an optimum growth temperature of 22-24°C and a water requirement of in excess of 2 mm per day for optimum growth, tick the appropriate box as to how you think growth would be affected by the following conditions.

Environment conditions (temp.) (water supply)	optimal	sub-optimal	very sub-optimal
25°C, 1.5 mm/day			
23°C, 2.5 mm/day			
15°C, 2.5 mm/day			
10°C, 0.5 mm/day			

This question is a little subjective because the exact impact that environmental conditions may have on a crop depends to some extent on the variety involved. The sort of pattern you should have recorded is:

Environment conditions (temp.) (water supply)	optimal	sub-optimal	very sub-optimal
25°C, 1.5 mm/day		X	
23°C, 2.5 mm/day	X		
15°C, 2.5 mm/day		X	
10°C, 0.5 mm/day			X

You might have judged 25°C and 1.5 mm/day as being very sub-optimal, since both parameters are outside the optimal range and they exacerbate each other. Relatively speaking, however, 10°C and 0.5 mm/day are much more obviously very sub-optimal.

When all the environmental factors are at optimal levels, maximal yields (within the genetic capability of the crop variety) can be obtained. Where one or more environmental factor is at sub-optimal levels, the crop is unable to develop at its maximal rate, and yields will be reduced if such conditions exist for any length of time. When an environmental factor severely reduces the growth rate of a crop, this is generally called a stress condition.

stress conditions

Obviously, a whole range of factors act on crops to determine final yield. Therefore, it becomes necessary to try to classify these factors into broad categories. The first major division made is between biological and physico-chemical factors. These are often called biotic and abiotic environmental factors, respectively. Biotic factors are those conditions surrounding the crop that affect it via some biological property - for example pests, diseases, or competition for light and nutrients by other plants within the crop stand. Biotic factors will be discussed in the next chapter. Abiotic factors are physical and chemical factors.

biotic and abiotic factors

3.2.2 Abiotic factors mediating crop productivity

Abiotic factors are generally physical or chemical factors acting on a plant or crop within its surrounding environment. Abiotic factors can be sub-divided into those that occur and act within the aerial parts of the crop. These are generally atmospheric in nature, and those that act in the soil. The most important of these are summarised in Table3.1.

abiotic factors of the atmosphere and soil

Aerial factors		Soil factors		
radiant energy		temperature		
light	- day length	soil moisture		
	- wavelength			
	- irradiance			
		soil atmosphere		
humidity		nutrient	-	concentration
temperature			-	availability
wind speed		pH		
cloud, mist, fog		soil structure		
rain and other precipitations		pollution		
atmospheric composition	- CO_2			
	- pollution			

Table 3.1 Abiotic factors affecting crop productivity.

The table is not exhaustive, but gives the main yield-affecting factors. There is a large number of abiotic factors and there could be a multitude of interactions between them.

Many of the environmental factors associated with the aerial sub-division are associated with climate and weather, and this is a theme that will be expanded as the chapter progresses.

climate and weather

3.2.3 The nature of crop production - the management variable

It must be remembered that, to the farmer, the objective of crop production is largely economic. In subsistence agriculture the objective is usually to feed a family or village group. In Western agriculture, it is to sell the harvested crop for a profit, meaning that the value of the useful yield must exceed the added costs of all the inputs (labour, machinery, fertilisers etc). The initial decision as to the choice of crop to grow, and the management of that crop throughout its life-cycle to harvest, act together with the environmental factors encountered by the crop during its life-cycle to determine the success of the farmer's enterprise; this is true in modern or subsistence agriculture. This is summarised in Figure 3.1.

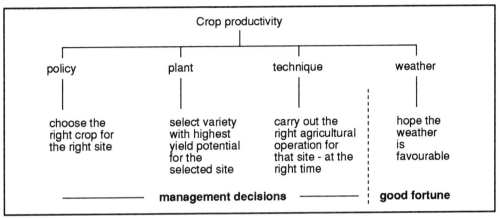

Figure 3.1 The influence of management decisions and good fortune on the outcome of a farming enterprise.

In short, a farmer must select a suitable crop that is well adapted to the climate of the region in which his farm is situated. It must be the variety with the highest yield potential for his farm. Appropriate crop management must take place to overcome those adverse environmental conditions over which there is some control (for example, for poor soil nutrient status add fertiliser; for low soil water, irrigate; etc). It is then necessary to trust to luck that the weather is favourable during crop development to give a good yield. Of course, the farmer must use his experience of the weather conditions that are likely to be encountered in his locality, and this must influence his decision.

farming outcomes depend on good management and on good fortune

Obviously, it would be very difficult for the farmer, by himself, to make all the best management decisions. Fortunately for the farmer, agronomists and crop physiologists have worked out the most appropriate cropping strategies in most countries. Most national governments run agricultural advisory services (for example, ADAS in the UK). In developing countries throughout the world, much needed experimental work is being carried out by the agricultural research institutes of the UN's Food and Agricultural Organisation (FAO) to determine the best management practices for crops in these regions.

role of agricultural advisory services

3.2.4 Climate and weather

Π In everyday language we tend to use the terms climate and weather rather interchangeably, but they do have separate meanings and, in terms of crop production, the distinction is important. Before reading on, see if you can distinguish between them.

Here are definitions of climate and weather:

• climate: refers to the broad-scale meteorological conditions that are characteristic of extensive land areas from season-to-season and over long time periods;

- weather: refers to short-term fluctuations in meteorological conditions within local areas, from day-to-day and hour-to-hour.

For the following meteorological conditions, distinguish between those that would be termed climatic conditions and those which would be termed weather conditions. Circle the appropriate answer.

1) A mean annual rainfall in excess of 45 cm allows cereal production in Western Europe (climate/weather).

2) 4 cm of rain on the evening of 24 June improved cereal yields in the Brussels area by an estimated 7% (climate/weather).

3) June rainfall of 8 cm can improve cereal yields in the Northern European Plain by up to an estimated 9% (climate/weather).

4) Winter temperatures of less than -18°C prevent the cultivation of winter wheat in the Canadian province of Ontario (climate/weather).

climate
influences the
choice of crop

Climate (the broad-scale meteorological conditions prevalent from year-to-year) determines what crops can be grown in a given location. For example, the cultivation of maize for sweet corn consumption in Northern Europe is restricted to a few sheltered, warm sites because, in general, climatic conditions are too cool to allow the ripening of cobs. The farmer would normally choose crops better adapted to his climatic conditions.

Although climate may determine the potential yield of the crop on a broad scale, it is the meteorological conditions during the few weeks or months of crop growth that determine its actual yield. These are, of course, the weather conditions met by the crop and they represent the short-term vagaries of climate. For example, the yields of a cereal crop on a small number of localised farms may be reduced for one year because June rainfall was only 20% of that normally experienced.

The importance of weather variations in determining yield are emphasised in Figure 3.2. It also reaffirms the relationship between potential yield, actual yield, good management and good weather (good fortune).

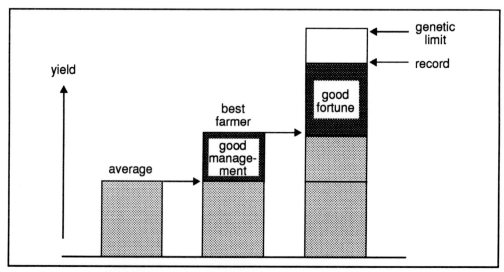

Figure 3.2 Stylised relationship between good management and 'good luck' on yields (see text). (Adapted from Tiry Agricultural Ecology, Longman Sci & Tech, Harlow).

The genetic limit is never met, of course, in outdoor-grown crops and, generally, yields are higher on experimental husbandry farms than on normal ones. However, yields from experimental farms set an obvious challenge to farmers.

3.2.5 Climatic control of crop distribution

In the previous section, the idea has been advanced that climate determines what crops can economically be grown in a given area. We must now examine which climatic factors are the most important.

In the vast majority of cases, it has been found that the distribution of vegetation on a global basis is determined by temperature and rainfall. Notice that the word vegetation has been used. Vegetation refers to the predominant plant community present in a given location, and the principles that determine vegetational distribution also apply largely to crop distribution. In Table 3.2 we have linked the predominant climatic groupings with the distribution of vegetational types.

Major climatic type	Chief characteristic	Sub-climatic/ vegetational type
tropical rainy	coldest month must be above 18°C	tropical rainforest monsoon rainforest tropical savannah
dry	evaporation exceeds precipitation	desert (hot, dry) Steppe (cold, dry)
humid, mild winter, temperate	coldest month is above 0°C but below 18°C	mediterranean humid sub-tropical marine west-coast
humid, severe winter, temperate	coldest month below 0°C, warmest above 10°C	humid continental (warm summer) humid continental (cool summer) sub-arctic
polar	warmest month below 0°C	tundra ice cap

Table 3.2 Relationship between predominant climatic types and vegetation types.

importance of temperature and precipitation

It is easy to see that, in the major subdivisions of vegetation, temperature and precipitation are the most important determinants of vegetation type. Crops tend to be distributed in a similar fashion, although one has to remember that a crop will only be grown at a given locality if it gives an economic yield. A good example of the distribution of major crops being influenced by climatic factors is given by the distribution of the world fruit crops shown in Table 3.3.

tropical	sub-tropical	temperate		
		mild winter		severe winter
coconut banana mango pineapple cacao	coffee date fig avocado	pomegranite citrus olive	almond blackberry grape	raspberry peach cherry apricot strawberry pear plum apple currant
low temperature sensitive	slightly frost tolerant	tender		winter-hardy

Table 3.3 Distribution of the fruit crops according to climate.

∏ From your own experience you should be able to cite many examples of this type of distribution. Think, for example, of the climate in Spain and then list the type of fruits that are grown there. You might like to do this for several different localities. For example, where does your climate fit on the scale shown in Table 3.3? Are the locally grown crops consistent with those indicated in this Table?

We would be surprised if your answer was no! Typically in European countries bordering the North Sea, we would anticipate finding fruit crops typical of temperate climate. These would include apples, pears and plums. In warmer areas, we would expect to find cherries, apricots, raspberries and peaches, whilst in still warmer areas we would anticipate finding grapes and almonds. In contrast, in very tropical regions such as Jamaica, we would expect to find crops such as coconuts, bananas and pineapples.

So overriding considerations in selecting suitable crops to grow in a particular region are the temperature and rainfall.

monitoring of meteorological conditions

Fortunately, temperature and precipitation are relatively easy to measure and assess. Throughout the world, they are continuously monitored at meteorological (frequently called weather) stations. These recordings are analysed by climatologists to give the following statistics:

temperature

absolute maximum during year
absolute minimum during year

mean daily and yearly temperatures
soil temperature

precipitation
(rain or snow)

mean annual precipitation (summer mean)
(winter mean)

mean monthly rainfall
mean seasonal rainfall

Obviously, these figures do change from year-to-year as local variations in weather occur, but over a period of several years there will be considerable stability in the overall statistics. It is only when a marked change can be seen in a climatic factor over many years that a change in climate is deemed to have occurred. This may then have an effect on crop distribution and crop yields.

long-term climate changes A well publicised example of long-term climate change is the reduction in rainfall in the Sahel (the areas peripheral to the Sahara desert). The reduction of water availability has gradually reduced crop growth and pasture production, leading to a reduction in the numbers of stock animals that the land can hold, thus reducing food availability to the indigenous peoples. This simple change in a climatic factor has led to widescale and widely publicised social consequences.

short-term extremes in weather conditions may greatly influence crop yields Although in general, temperature and precipitation are the most important factors determining crop distribution worldwide, in fact, it is often the single environmental factor that is least favourable to plant growth at any point within the growth cycle of a plant or crop species that determines whether that plant can grow at a given location. Often, it is the extremes of climate rather than small subtle day-to-day changes in climate that determine distribution. Periodic winter cold or periodic drought may be much more crop-limiting than mean annual figures for temperature or rainfall. For example, although mean summer temperatures in Canada would allow cultivation of winter wheat, periods of extremely low temperatures in winter tend to kill the winter variety which makes the spring wheat a preferred crop.

A useful concept in discussing climate and crop growth is the concept of day-degrees. In temperate regions plant growth virtually ceases if the temperature falls below 6°C. When temperatures are below this, there is no plant growth. If, however, the temperature rises above 6°C, plants begin to grow. The rate of growth is, over the temperature range encounted in temperate regions, proportional to the temperature. We can write the relationship:

growth rate = $k(T-6°)$

where T = temperature and k = constant.

Now let us consider a daily cycle in temperature. We can represent this in the following way:

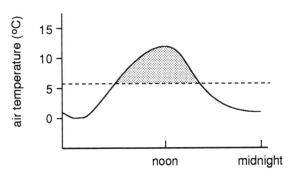

daily cycle of air temperature

The area enclosed by the curve and the line drawn at 6°C gives a quantitative indication for growth.

∏ What units does this area have?

The answer is day (or hours) degrees.

∏ Will the average day-degree per week be the same throughout the year in Europe?

The answer is clearly no. As the temperature rises during the summer, so will the day-degrees.

For example, in the UK, the day-degrees accumulated per week in January will be of the order of 2-4 whilst in July this will rise perhaps as high as 75-80.

use of the concept of day-degrees to predict harvesting time
This system enables growers to plan sowing operations and to predict harvesting times. Say, for example, it is known that 1000 day-degrees are the minimum that are required to produce a commercially accepted yield of carrots. By using data available from meterological services concerning temperatures, it becomes feasible to calculate when a crop might be ready for harvesting.

This approach is, of course, rather a simplification since it assumes that nutrients and drought will not restrict growth. It also assumes that for any 1° rise in temperature it has the same effect on growth providing it is above 6°C. This is a reasonable assumption providing the temperature does not rise too high. At high temperatures (for example 35°C) plant growth ceases. In practice, the relationship we described above appears to be a reasonable one in the range 6-30°C, which covers most circumstances in temperate regions.

3.3 Microclimate and the crop canopy

In Section 3.2 the broad-scale effects of environmental factors, as determined by both climate and weather, have been outlined. So far, climate and weather have been approached without reference to any modifying effects that the plant or crop could exert.

To the crop physiologist, it is the relationships between plants growing together and their combined effects on the overall physiology of the plants as a group, that is of interest, rather than the physiology of a single plant. This is because plants growing together alter the environmental factors acting in their immediate proximity, which also has an effect on their overall physiology and yield response. When plants are growing in large numbers as a crop stand, the plant mass is known as the crop canopy and the immediate environment surrounding, and within, the crop canopy is known as the crop microclimate or microenvironment.

crop canopy and microclimate

Figure 3.3 gives some actual data for a range of environmental parameters measured both above and within a maize crop. Parameters have been measured at different heights through the crop, and, at sites where the thickness of the crop differs.

Remember that maize is a tall crop, in this case the top of the canopy is about 220 cm above ground level. Note that the environmental factors within the crop canopy are rather different from those above the canopy. In some cases (eg wind speed), the crop canopy exerts quite an effect immediately above itself. To analyse what happens to the range of environmental factors within the canopy, an estimate of how plants are arranged within the canopy is required. In the following discussion we will refer to a

leaf area density — crop with a total leaf area density of 3.5 $m^2 m^{-3}$ (continuous lines in Figure 3.3; the dotted line is referred to later in the text). Leaf area density (LAD) is often used as a measure of the crop biomass in space because the leaves represent the assimilatory (photosynthetic) component of the plant/crop which produces biomass. LAD is the ratio between leaf area and volume of space occupied by the plants. In this case total LAD is 3.5 m^2 for every 1 m^3 occupied by the crop.

In Figure 3.3 you will notice that LAD has been reported at various heights through the crop. Note that LAD is maximal at about 120 cm above the ground. By reporting LAD in this way it shows how the leaves are profiled throughout the canopy.

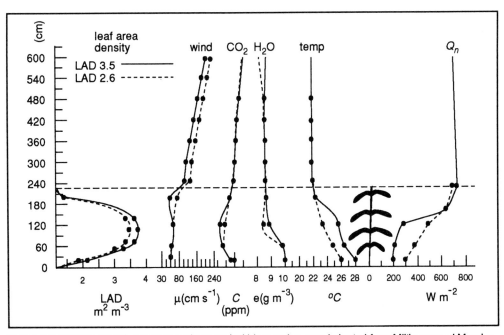

Figure 3.3 Environmental parameters above and within a maize crop (adapted from Milthorpe and Moorby, An Introduction to Crop Physiology, Cup.). (see text for further explanation). μ = windspeed and is reported in cm s^{-1}; C = concentration of CO_2 in ppm; e = humidity in g m^{-3}; Q = radiation energy per time per unit area (W m^{-2}).

Environmental factors alter within the canopy in the following fashion:

wind speed (μ) is reduced both above and within the canopy compared with the wind speed at a height of 600 cm above the canopy. Contact with a solid surface (ie a leaf) has the effect of reducing particle velocity, with wind being most reduced at ground level. Wind has an effect on any movement of matter (for example, water vapour, CO_2) within the canopy, and also exerts a cooling effect;

• CO_2 levels (C) are uniform above the canopy but are reduced within the canopy. This reduction is of course greatest where photosynthesis is maximal (ie at maximal LAD). Any shading of leaves by other leaves, reducing photosynthesis, will also affect CO_2 levels;

- H_2O levels (e) rise within the canopy. Raised H_2O levels are the result of the evaporation of water from leaves. Transpiration (ie water loss via the stomata) is affected by the energy absorbed by the leaf, leaf temperature and removal of water vapour by wind;

- temperature (°C) rises several degrees in the canopy and is maximal at ground level. Temperature is affected by the radiation input (energy absorbed as light energy from the sun), and any cooling effect due to the wind and evapo-transpiration;

- net radiation (Q_n) is progressively reduced through the canopy to ground level. Leaves absorb solar radiation - to the greatest extent where leaf density is maximal. Net radiation is described in greater detail later in this chapter.

| SAQ 3.2 | Using Figure 3.3 see if you can describe what happens to the various parameters measured when leaf growth has been less extensive and produced a LAD of $2.6 \ m^2 \ m^{-3}$ (dotted line in the figure). |

From SAQ 3.2 and the example described, it is fairly clear that the interactions between the crop canopy and environmental factors are rather complex. What may appear as a simple change in one factor can lead to widespread changes in the crop microclimate, which can have profound effects on crop physiology, which in turn, can lead to a changed crop yield.

3.4 Environmental factors affecting crop physiology

From the previous section, you can see that a whole range of individual environmental factors is involved in the microclimate in the crop canopy. Individual factors interact to bring about the overall physiological response of the crop, which results in the final yield of harvestable product. The complexity of these interactions has led crop physiologists to try to assess the effects of factors individually, in order for us to be able to understand the role of each factor in determining physiology and yield. Obviously this is a complicated and lengthy subject and many textbooks and academic symposia are devoted to it. In the following sections, case studies are given for some of the most important factors that control crop yield. An example of a factor within the aerial environment has been chosen, namely the effect of light on crop growth, and also the effects of two important soil factors, the availability of nutrients and of water, are dealt with later.

3.5 The effect of light on crop growth and yield

3.5.1 The biophysics of light

Light represents that portion of the electromagnetic spectrum that we can see, ie those wavelengths that are visible to the naked eye. Strictly speaking, the origin of light and most of the electromagnetic spectrum can be ascribed to the activity of the sun.

solar radiation   Generally that portion of the spectrum that reaches the Earth's surface is known as solar radiation. It is solar radiation rather than light that has effects on plant growth and development. The Earth's atmosphere filters out much of the more harmful high energy radiation that hits the Earth from space, (especially in the ultraviolet region). Figure 3.4 shows the different types of radiation found within the electromagnetic spectrum, and gives both the wavelengths and frequencies of the various forms.

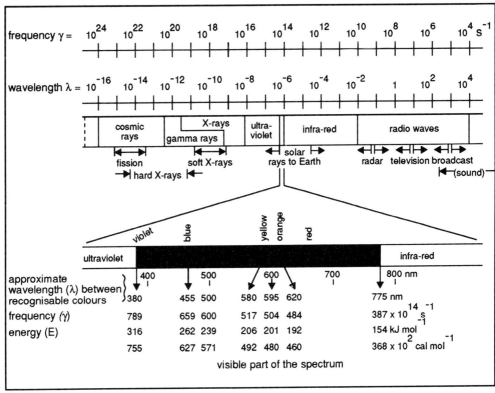

Figure 3.4 Electromagnetic spectrum showing the energy of radiation in the visible part of the spectrum. (Adapted from Salisbury and Ross, Plant Physiology, Wadsworth, Belmont Co, 1992).

Note that the harmful high-energy radiation is found on the left-hand side of the figure and is characterised by short wavelength and high frequency. Note also that the solar radiation reaching the Earth represents a narrow band of the whole spectrum, ranging from the UV, through the visible to the infra-red. The breakdown of the components of the visible region is shown on the expanded scale.

∏   Examine the scale of wavelengths and frequencies of the visible region. Which coloured light has the highest energy?

Energy of the different forms of light is lower at longer wavelengths and lower frequency. Therefore, violet light (400 nm) is the most energetic and red light (greater than 620 nm) is the least energetic. You can also note the wavelength range of solar radiation from the diagram.

Although we have looked at radiations (such as light or solar radiation) as a wave form, physicists look at radiation not only as a wave form but also as discrete packets of energy (rather like a stream of energetic particles). In either case, you must remember that all radiation has an energy content. If we think briefly in terms of energy packets, the smallest possible packet of radiant energy is known as the quantum (plural = quanta). At a particular wavelength range such a packet of energy is also called a photon, and it is this quantity that is largely used by crop physiologists interested in the effects of radiant energy. In basic terms, a single quantum or photon, if absorbed, will cause the ejection of a single electron from the absorbing surface.

quantum and photons

It is important that we can calculate the amount of energy (E) in a photon of radiant energy (light), as the energy content can cause certain photochemical reactions within the plant, such as photosynthesis, to occur. This relationship is given by the expression:

$$E = \frac{hc}{\lambda}$$

where h = Planck's constant, c = velocity of light and λ = wavelength.

Frequency and wavelength are related through the velocity of the wave accordingly:

$$frequency = \frac{velocity}{wavelength}$$

So far, we have discussed the energy of a single photon, which is so small as not to be of great use to the crop physiologist who needs to consider large-scale photochemical processes. In these circumstances, it is more meaningful to calculate the energy content of a mole of photons (1mole = 6.02×10^{23} particles).

SAQ 3.3

Using the previous equations, calculate the total energy in a mole of red light photons. (wavelength = 660 nm, c = 3×10^8 m s^{-1}, and Planck's constant = 6.63 x 30^{-34} J s^{-1}).

Hint: first of all, calculate the energy in a single photon and then multiply by the mole number to get the molar quantity.

What units are used?

radiant energy flux and irradiance

Obviously, plants do not receive energy at a single wavelength, but across a wide spectral range. Therefore, the total energy received by a surface is of more relevance than the energy received at a single wavelength. Two parameters of value to the crop physiologist are radiant energy flux and irradiance:

- radiant energy flux is the amount of radiant energy (joules, J) falling on a surface per unit of time. Units are: J s^{-1};

- irradiance is the amount of energy received by unit surface area per unit time. Units are: J s^{-1} m^{-2}.

There are alternative ways of expressing irradiance; as W m^{-2} (since 1W = 1J s^{-1}), or on a photon basis (μmol photons m^{-2} s^{-1}).

3.5.2 Solar radiation and plant processes

A leaf can be considered as a surface receiving an energy input from solar radiation. A surface such as a leaf, as well as absorbing radiation, also emits radiation, especially in the infra-red region of the spectrum. Solid bodies such as a leaf emit radiation according to the Stefan-Boltzman law, which states that all objects above a temperature of absolute zero emit radiant energy.

The fate of radiation that meets a leaf surface is threefold. Radiation that is not absorbed by the leaf is reflected or is transmitted through the leaf. Radiation that is not reflected or transmitted will be absorbed by pigments. Some of the absorbed radiation may be emitted (ie re-radiated). The net radiation is the difference between absorption and emission, and represents the amount of energy absorbed and retained by the leaf.

some radiation energy lost as heat

Radiation losses can also be due to energy transfer processes. Radiation absorption will lead to a rise in the energy status of the leaf and, therefore, its temperature. The heat can be lost from the leaf by convection, conduction and also by transpiration. Transpiration, in particular, represents the major pathway by which heat is lost from the leaf.

A pigment is a substance that is able to absorb radiation of a particular wavelength. Pigments are only able to absorb certain wavelengths, and this is certainly true for pigments contained in plant leaves. In Figure 3.5 we show the relative absorptive properties of the pigments found in leaves and involved in photosynthesis.

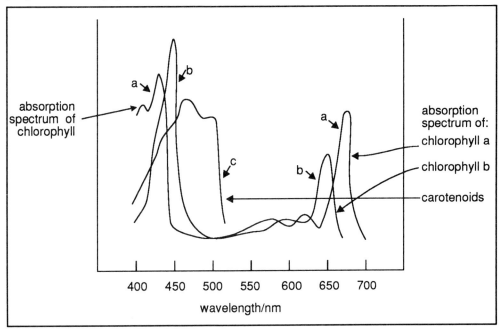

Figure 3.5 Absorption spectrum of chlorophyll a (a), chlorophyll b (b) and absorption spectrum of β-carotene (c).

You can see from Figure 3.5 that chlorophylls absorb strongly in the 400-500 nm and the 600-700 nm regions, and the carotenoid accessory pigments absorb between 390-510 nm range. Within these ranges each pigment has characteristic absorption peaks. Bodies which completely absorb incoming radiation will not reflect any of that radiation, and therefore appear black. Leaves do not absorb (ie they reflect and transmit) radiation between 500 and 600 nm, which is the green part of the visible spectrum. Therefore, leaves appear green.

Also from Figure 3.5 you can see that absorption occurs within the wavelength range 400-700 nm. It is only radiant energy in this range that can be used for photosynthetic reactions, and consequently crop physiologists need to know the amount of energy available in this region of the spectrum. Sensors have been developed to measure radiant energy in the region 400-700 nm, which is known as **photosynthetically active radiation (PAR)**.

photosynthetic-ally active radiation

Photosynthesis, of course, is a photochemical process - a series of chemical reactions driven by the energy from light (solar radiation). Physiologists need to know how effective solar radiation is in driving these chemical reactions. The **quantum yield** is a measure of efficiency according to the following equation:

quantum yield

$$\theta = \frac{M}{q}$$

where θ = quantum yield, M = number of molecules reacted and q = number of photons absorbed.

photosynthetic
photon flux

You can see from this equation that photosynthesis is not dependent on the amount of energy absorbed but on the number of photons absorbed. Despite their different energy levels a red photon has the same effect on photosynthesis as a blue photon. In photosynthesis it is, of course, photons within the range 400-700 nm that are effective, and crop physiologists now commonly use the term the photosynthetic photon flux to describe the number of photons (400-700 nm) available to the leaf per unit area and time (μmol m^{-2} s^{-1}).

∏ We have stated that photosynthesis is not dependent on the amount of energy absorbed but on the number of photons absorbed. Bearing in mind the absorption spectra of chlorophylls and carotenoids such as β-carotene, should you anticipate that the amount of photosynthesis will be dependent not only on the amount of light irradiating a plant, but also on its wavelength? If you conclude that it will be dependent upon wavelength, give your reasons.

The answer is that the amount of photosynthesis is dependent upon the wavelength as well as on the intensity of radiation. Since plant photosynthetic pigments do not absorb light to the same extent at all wavelengths (see Figure 3.5), the number of photons absorbed per number of incident photons also depends upon wavelength.

In Figure 3.6, we have re-drawn the absorption spectrum of a plant and drawn onto it an 'action spectrum' of photosynthesis.

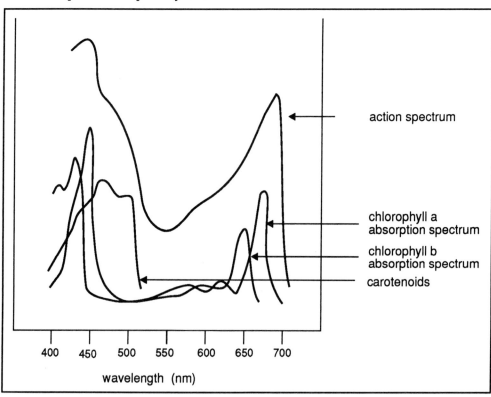

Figure 3.6 Adsorption spectra of plant pigments and the action spectrum of photosynthesis. Note that the action spectrum follows the general profile of the spectra of the pigments.

There is quite close agreement between the action spectrum and the absorption spectrum. They are not, however, identical. Why is this so?

some
light-absorbing
components
are not
involved in
photosynthesis

You should realise that plants contain low levels of light-absorbing components that are not directly involved in photosynthesis. That is, they cannot pass the absorbed energy onto the energy-trapping mechanisms involved in photosynthesis. Furthermore, photosynthesis in higher plants involves two separate photosynthetic events involving two photosynthetic centres (called photosystems I and II). These two centres work in unison but each absorbs slightly different wavelengths of light. Since both systems are needed for the proper functioning of photosynthesis, the relationship between absorption and photosynthetic activity is rather complex.

Details of the physiological and energetic aspects of higher plant photosynthesis are given in the companion BIOTOL text, 'Crop Physiology', so we will not go into specific details here. For our purposes, it is sufficient to regard sunlight as 'white light' (= light containing photons of all the wavelengths of the visible spectrum) and to remember that the photosynthetic action spectrum is very similar to the absorption spectrum of the plant.

3.5.3 The effect of solar radiation on crop yield

During the course of a growing season a finite amount of PAR is incident upon a unit of ground area (eg a hectare) and, therefore, available to a growing crop. The amount of PAR that is incident on a particular site varies due to changes in day length and seasonality. In temperate regions, maximal incidence occurs during the summer months, and is markedly affected by latitude. The total amount of radiation is a function of irradiance and duration (day length).

Let us examine the availability of light in a little more detail. The radiation flux at the outer edge of the atmosphere of the Earth, measured perpendicularly to the sun, is almost constant and equal to $1370 \text{ J m}^{-2} \text{ s}^{-1}$. This is known as the solar constant and is usually symbolised by S. The incoming radiation, however, depends on the angle at which the sun's rays strike the atmosphere. At lower elevations, there is a corresponding lower radiation flux per unit area. The radiation which actually reaches the outer edge of the atmosphere is denoted by the symbol S_0. The elevation of the sun is, of course, a function of latitude, day of the year (season) and time of day. Thus, the intensity of light, in $\text{J m}^{-2} \text{ s}^{-1}$, at the surface of the Earth depends on latitude, day of the year and time of the day.

solar constant

Π What else influences the light intensity reaching the Earth's surface?

global
radiation

The factors we were looking for are the atmospheric conditions such as clouds and dust. These determine how much light is absorbed and reflected on its way to the surface. We have illustrated this in Figure 3.7. The radiation that reaches the Earth's surface is called global radiation (Sg).

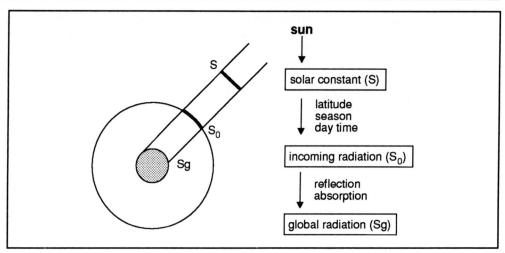

Figure 3.7 Relationship between solar constant, incoming radiation and global radiation.

The radiation that is available for plants is, of course, the global radiation (Sg).

The effects of latitude and time of the year on the global radiation are illustrated in Figure 3.8. The data in this figure assume that no absorption or reflection has taken place in the atmosphere and is reported in MJ m^{-2} day^{-1}.

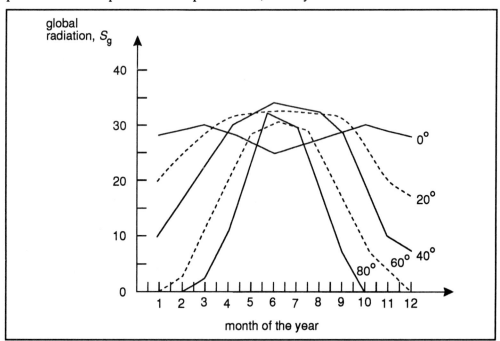

Figure 3.8 Daily global radiation on clear days as a function of latitude and time of the year.

Ⅱ Which latitude illustrated in Figure 3.8 shows the greatest variation in daily global radiation during a year?

You should have identified a latitude of 80°.

Day length also varies with latitude and the time of year (Figure 3.9).

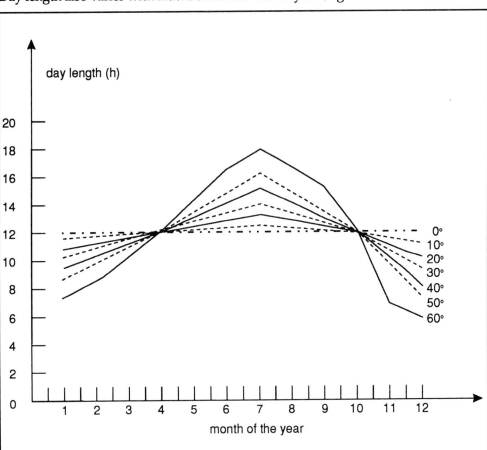

Figure 3.9 Length of day as a function of latitude and time of the year.

Dust and water vapour in the atmosphere absorb and scatter radiation before it reaches the Earth's surface. Thus the actual value of Sg varies from day-to-day and consists partly of direct radiation and partly of scattered (diffused) radiation.

atmospheric transmission coefficient

The ratio between the global radiation at the Earth's surface and at the edge of the atmosphere (S_g/S_0) is defined as the atmospheric transmission coefficient. On an overcast (cloudy) day the transmission coefficient is about 0.2 while on a bright, clear day it is about 0.8. Typical average values over the growing season in, for example, the Netherlands (52°N) is about 0.42, whilst in Kenya (0°N) is nearer 0.54.

In the absence of a direct measure of daily global radiation, PAR is used.

Π   What happens to the global radiation (or PAR) at the leaf canopy?

We learnt earlier that some is reflected, some absorbed and some may pass through the leaves (transmitted).

We can illustrate this in the following way:

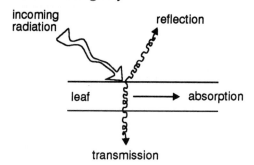

It should be self-evident that if the leaf canopy is more than one leaf thick, then the incident (incoming) radiation for the second layer will be the radiation that is transmitted through the upper layer.

Some data relating to incident and absorbed PAR and CO_2 assimilation per leaf layer are shown in Table 3.4. Note that the incident PAR is reduced at each layer and that CO_2 assimilation is also reduced in lower layers. However, the efficiency of light use, in terms of CO_2 assimilated per unit of incident light energy absorbed, rises at the lower light intensities.

leaf layer	incident PAR $(J\,m^{-2}\,s^{-1})$	absorbed PAR $(J\,m^{-2}\,s^{-1})$	assimilation $(kg\,CO_2\,ha^{-1}\,h^{-1})$	light use efficiently $(kg\,CO_2\,ha^{-1}\,h^{-1}/J\,m^{-2}\,s^{-1})$
leaf layer 1	100	80	23.74	0.30
leaf layer 2	10	8	3.44	0.43
leaf layer 3	1	0.8	0.36	0.45

Table 3.4 Incident radiation, absorbed PAR and CO_2 assimilation per leaf layer in a crop with three horizontally extended leaf layers. (Incident PAR above crop canopy = 100 J m^{-2} s^{-1}).

∏ See if you can explain the differences in light use efficiency reported in Table 3.4.

Your ability to answer this will depend upon your knowledge of plant physiology. Leaves of C_3 plants become light-saturated at PAR values of the order of $250\,J\,m^{-2}\,s^{-1}$. As this value is approached, the use of absorbed light in terms of CO_2 assimilation becomes more inefficient.

∏ What would happen to the light use efficiency if the incident PAR was raised to $500\,J\,m^{-2}\,s^{-1}$?

Raising the light intensity to this level would lead to only a modest increase in the assimilation rate. This would result in a lower light efficiency, especially in the first leaf layer. In these circumstances, however, more radiation would penetrate to the lower leaf layers and so the amount of CO_2 these would assimilate, would be increased.

The relationship between incident PAR and assimilation is, therefore, complex. In spite of fluctuations in the incident light over a growing season, there is usually quite a reasonable correlation between biomass production and the amount of intercepted radiation. We have stylised this relationship in Figure 3.10.

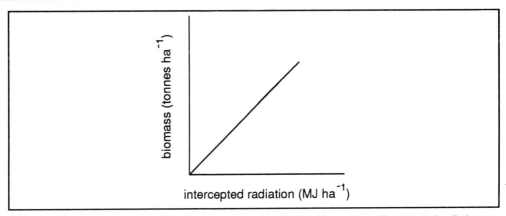

Figure 3.10 Empirical relationship between biomass production and the amount of intercepted radiation.

SAQ 3.4

Draw a graph similar to that shown in Figure 3.10 to demonstrate how plants with higher and lower levels of photosynthetic efficiency would accumulate biomass at the given ranges of intercepted radiation.

Biomass production in response to actual quantities of intercepted radiation has been determined for several crop species, and some examples are given in Figure 3.11. Note that there is a linear relationship in all three examples given. This has led many crop physiologists to the conclusion that photochemical efficiency is constant through a growing season once canopy development is complete and that crop photosynthesis is independent of temperature over the range 10-25°C. This suggests that solar radiation is the single most important factor determining crop yield, assuming that water and nutrients are not limiting.

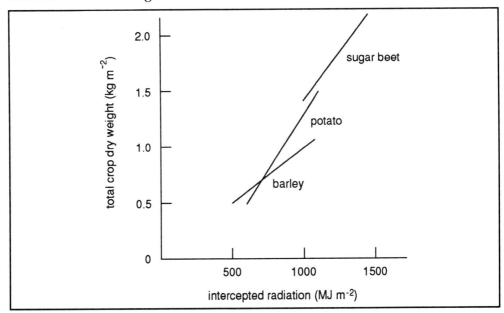

Figure 3.11 Total crop production and intercepted radiation for three major European crops.

The amount of PAR absorbed depends largely on the architecture (ie structure) of the crop canopy and the presence of that canopy at the right time to maximise the availability of solar radiation. As discussed earlier, crop canopies are complex structures, often with complex arrangements of leaves. It is the leaves which absorb radiation leading to accumulation of biomass. A convenient way to express the extent of the crop canopy is the leaf area index (LAI), which is the ratio of leaf area to ground area. For many crops, the relationship between LAI and the amount of incident radiation has been worked out. We illustrate this in Figure 3.2 using data for various potato crops.

leaf area index

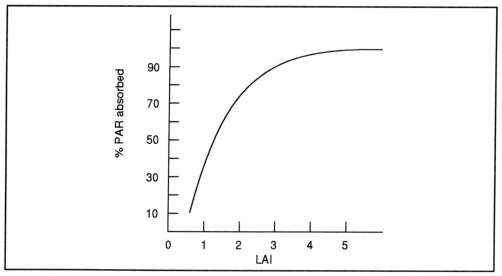

Figure 3.12 Relationship between %PAR absorbed and LAI for potato crops (see text for details).

∏ At what LAI would maximal biomass accumulation take place?

Your answer should be a LAI of about 3.5 - 4. Above this value there is virtually no increase in light absorbance and thus no increase in photosynthesis. It was originally thought that above an optimal LAI (about 3.5 in the potato example) leaves lower down in the canopy were being shaded by the higher leaves, and would, therefore, be near their light compensation points. We remind you that the light compensation point is the level of light intensity where the amount of CO_2 fixed by photosynthesis is equal to that lost due to respiration. This could mean that the shaded lower leaves could be a drain on assimilates fixed by the upper leaves and would diminish yield if they were below the light compensation point. This, of course, is in contrast to thinner canopies where LAI is low and plants in the canopy too widely spaced to absorb all the incident radiation. In this case, maximum production per unit area could not be reached.

optimal LAI values

∏ From the data in Figure 3.13 on CO_2 fluxes (photosynthesis and respiration), do you think that high LAI is actually detrimental to yield?

The crop growth rate is effectively equivalent to the net amount of assimilate fixed (net photosynthetic rate = gross photosynthetic rate - respiration rate). This does not decline at high LAI. At low LAI, net photosynthetic rate is reduced. This has led to the concept of a critical LAI where crop growth rate is optimal. To maximise crop yield, LAI has to be held above a critical level. Although creating crop stands with canopies with LAIs much above this level would waste seed and reduce maximal profitability.

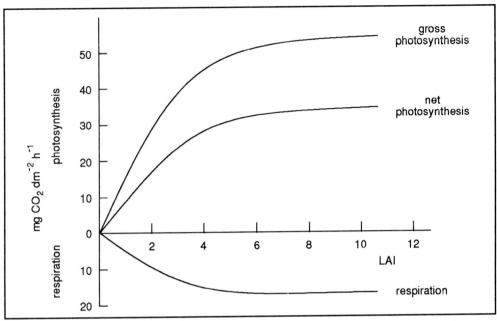

Figure 3.13 Relationship between gross photosynthesis, net photosynthesis and respiration rates for wheat crops of different LAI values (incident light = 400 J m^{-2} s^{-1}).

We have seen so far that the relationship between incident radiation and net assimilation is dependent upon LAI. You should also anticipate that the relationship between biomass production and incident PAR will be dependent on the morphology of the leaves, especially the angle at which they are held relative to the ground and the incident light.

SAQ 3.5

Below is a graph of gross CO_2 assimilation rates against incident PAR for a crop with an LAI = 5 and one with an LAI = 1.

Explain why the LAI = 5 curve is not 5x that of the LAI = 1 curve?

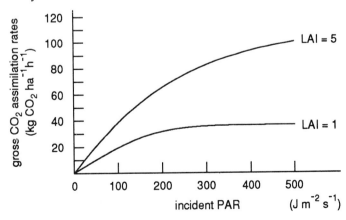

3.6 The effect of nutrients and water on crop growth

In the previous section we looked at the effect of light (or rather solar radiation) on crop growth and yield. Crop yield can only be maximised in relation to intercepted radiation if soil nutrients and water are present at sufficient levels. In the following sections, we will look at the effects of soil nutrients and water on crop growth, and consider the potential consequences when levels are sub- or supra-optimal.

3.6.1 Soil nutrients

For optimal growth and development of a crop, the balance of mineral elements in the soil environment has to be at, or near, an optimal overall composition. If one or more of the required mineral elements (nutrients), are present at sub-optimal concentrations, then crop growth rates may be reduced. Such mineral elements are termed essential elements. Essential elements by definition have to fulfil all of the following criteria:

- plant growth cannot take place in their absence;

- they are required as a component of, or an activator of, a metabolic process;

- they cannot be replaced by another element.

Essential elements are divided into macro-elements, which are required at relatively high concentrations (present in dry plant tissue at above 0.1% of dry weight) and micro-elements (present below 0.1% of dry weight). Individual essential elements are given in Table 3.5 with typical amounts found in plant tissue.

Element	Chemical symbol	Form available to plants	Concentration in dry tissue mg/kg	Relative No. of atoms compared to molybdenum
molybdenum *	Mo	MoO_4^{2-}	0.1	1
nickel *	Ni	Ni^{2+}	0.4	5
copper *	Cu	Cu^+, Cu^{2+}	6	100
zinc *	Zn	Zn^{2+}	20	300
manganese *	Mn	Mn^{2+}	50	1000
boron *	B	H_3BO_3	20	2000
iron *	Fe	Fe^{3+}, Fe^{2+}	100	2000
chlorine *	Cl	Cl^-	100	3000
sulfur **	S	SO_4^{2-}	1000	30000
phosphorus **	P	$H_2PO_4^-, HPO_4^{2-}$	1000	60000
magnesium **	Mg	Mg^{2+}	2000	8000
calcium **	Ca	Ca^{2+}	5000	12500
potassium **	K	K^+	10000	250000
nitrogen **	N	NO_3^-, NH_4^+	15000	1000000
oxygen ***	O	O_2, H_2O	450000	30000000
carbon ***	C	CO_2	450000	35000000
hydrogen ***	H	H_2O	60000	60000000

Table 3.5 Mineral content of plants. Note these are typical values. Individual species may show some variations. Note: * = micro-elements; ** = macro-elements; *** = essential structural elements. Adapted from Salisbury and Ross, Plant Physiology, Wadsworth, Belmont Ca, 1992.

Where one or more of these elements is/are available at levels which are sub-optimal to the crop, the plants within the crop may show deficiency symptoms. Precise details and concentrations at which symptoms occur have been worked out for many crop species. In general, deficiencies are observed as growth reductions, changes in growth characteristics, characteristic leaf colorations, leaf chlorosis and leaf necrosis. The precise symptoms will vary with both the individual essential element, the soil concentration of the element, and the plant species. Figure 3.14 shows the generalised response of plant growth to increasing levels of a plant nutrient in the soil.

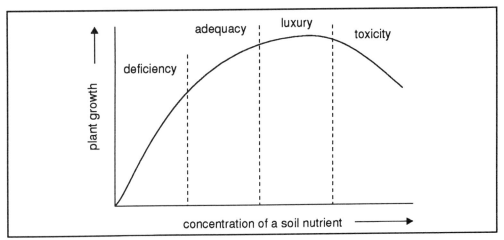

Figure 3.14 Generalised response of plant growth to increasing levels of a plant nutrient (assumes all other nutrients present in adequate amounts).

A whole range of essential elements will be required for optimal functioning of many metabolic processes. The absence of one element, even when all others are plentiful, will prevent functioning. This is illustrated in Table 3.6 which details the elements required in carbon-fixation, and gives the process in which each element has a role.

Process	Mineral nutrient involved	
	Structural compounds	Enzyme activators
chloroplast formation		
protein synthesis	N, S	Mg, Zn, Fe, K
chlorophyll synthesis	N, Mg	Fe
electron transport		
PSI, PSII, photophosphorylation	Mg, Fe, Cu, S,P	Mg, Mn
CO_2 fixation	-	Mg, K, Zn
stomatal movement	-	K, Cl
starch, sugar metabolism	P	Mg, P

Table 3.6 The role of mineral nutrients in carbon-fixation (adapted from Marschner, Mineral Nutrition of Higher Plants, Academic Press,1986).

Because metabolic processes are very complex a large number of elements is required at each stage in these processes. For each essential element, a yield response curve can be made. In general, these follow the forms shown in Figure 3.15, although each element will have its own curve.

∏   In Figure 3.15, two curves are given. Is element 2 likely to be a micronutrient or macronutrient?

You should have suggested that it was a macronutrient. Fairly large amounts (and, therefore, a substantial soil concentration) are needed to support growth. Element 1 is likely to be a micronutrient and required only in small amounts. Thus, a low soil concentration is sufficient to support growth.

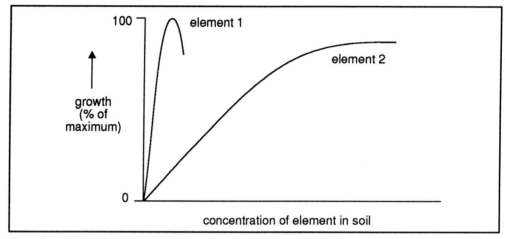

Figure 3.15 Forms of curves relating growth to soil concentration.

∏   What else does the curve for element 1 tell you?

toxicity of nutrients present in supra-optimal amounts

If the nutrient is present at sub-optimal concentration, a reduction in growth results. Similarly, some nutrients (particularly micronutrients such as Mo, Mn, Fe, Cu and B) can cause growth reduction if present in soils at supra-optimal concentrations. These are toxic conditions and for microelements, in general, they occur at low concentrations (in the µmolar to mmolar range).

The farmer is able to correct a deficiency of soil nutrients through the addition of fertilisers. After analysis of soils and a knowledge of a crop's requirements for nutrients, a policy for fertiliser application can be made for a given site. The main variables that must be considered are:

• crop species;

• the existing nutrient status of the soil;

• choice of fertiliser type;

• time of application.

We will discuss each in turn.

The crop species. Different crops respond differently to fertiliser applications. This is illustrated in Figure 3.16 for the application of nitrogenous fertiliser to beet, grass and cereals.

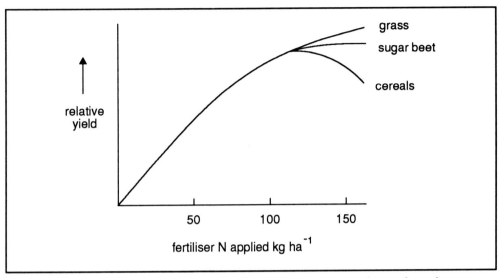

Figure 3.16 Effects of nitrogenous fertiliser on harvested yields of grass, sugar beet and cereals.

The existing nutrient status of the soil. If the general balance and concentration of the soil nutrients is depleted, application of a single nutrient element may not give the best yields, ie the yield response is limited by the deficiency of another element.

In Figure 3.17 we show the effects of applying a nitrogenous fertiliser on the yield of sugar beet, grown in soils with high and low phosphorus (phosphate) content.

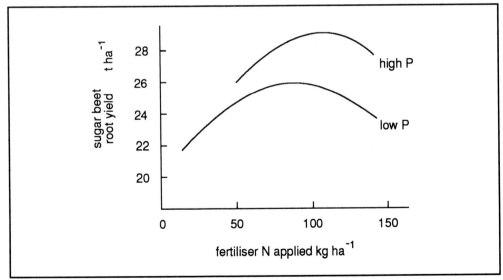

Figure 3.17 Effects of applying nitrogenous fertiliser on the yields of sugar beet in high or low phosphate-containing soils.

Note that the addition of phosphorus (as phosphate) to the soil improves sugar beet root yields, particularly when N-fertiliser is also added. This also demonstrates that there is an interaction between nutrient elements to achieve maximum yield.

The choice of fertiliser type. Fertilisers are applied either as straight fertilisers (which contain only a single useful plant nutrient) or as compound fertilisers (which contain 2 or more principal nutrients). In general, only macronutrients are applied regularly, most often as a composite of the nutrient elements N, P and K. It is important that the balance between the fertiliser components is correct for the crop being cultivated, to maximise yield.

Timing of application. The timing of application for a given nutrient depends on the mobility of the nutrient in the soil and on the growth cycle of the crop. For example, P and K are not very mobile within the soil horizons and their removal by rainfall (often called leaching losses) is low. In contrast, N is quickly lost by leaching, being very mobile in the soil horizons.

3.6.2 Nutrient cycling

So far, we have considered the nutrients within a soil to act as a simple static system. In fact, concentrations of nutrients within a soil are affected by a range of environmental factors, and can be held in more than one form (pool) within the soil. In general, the factors affecting the size and stability of a pool of a soil nutrient can be described by a nutrient cycle. Figure 3.18 gives an idealised example of such a cycle. Do note that there can be more than one pool of a nutrient within a soil, and that these different pools can be affected differently by environmental factors.

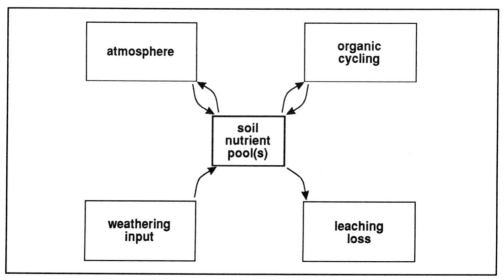

Figure 3.18 A simplified and generalised nutrient cycle (see text for further explanation).

The nutrient cycle represents a simple balance sheet describing the inputs and losses of a nutrient element within a soil system. The complexity of a cycle varies with the nutrient under consideration and a specific example of a nutrient cycle (the nitrogen cycle) is given in the next section.

3.6.3 Effect of a single soil nutrient on crop growth and yield - nitrogen

For most crop systems, soil (and hence crop) levels of the macronutrients, P, K, Ca, Mg, and S, can be easily manipulated by fertilisation so that they are held in the optimum range for crop growth. Nitrogen is the macronutrient required in greatest amount by the crop, being utilised in a wide range of metabolic processes and being found in a huge range of structural compounds within a cell. Hence, deficiency in soil nitrogen levels can lead to severe reductions in crop yield. This section will describe in detail how nitrogen is held within the soil environment and the effects it has on crop growth and yield. This section is intended to serve as an example of the complex interactions between a soil nutrient and the final yield in crop plants.

3.6.4 The effect of N on crop growth

Ⅱ From Figure 3.19 and Table 3.7, relate the responses of cereals and perennial rye grass to the availability of soil nitrogen.

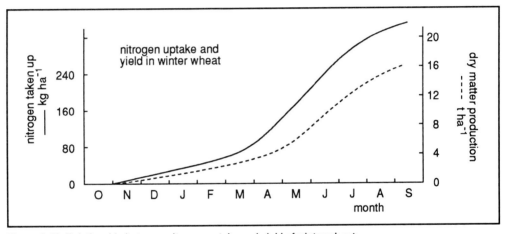

Figure 3.19 Relationship between nitrogen uptake and yield of winter wheat.

location	nitrogen fertiliser (kgN/ha/year)		
	(200)	(400)	(600)
S England	6.4	9.3	9.6
Midlands	5.6	9.6	10.3
N England	6.8	8.3	7.9
NW England	9.0	11.0	12.1
E Wales	6.3	9.8	11.2

Table 3.7 The effects of fertiliser on the yield of rye grass in five regions of the UK. (Data from Jackson and Williams, Agrichemicals, BCBP monographs, 1979).

From the data, you can see that N applied to the soil increases the yield of perennial rye grass, especially at lower application rates. When applied at higher levels, however, (eg 600 kgN ha^{-1} yr^{-1}) there is little extra yield from the higher application rate, and in some cases a reduction in yield is found. This, of course, relates to the general uptake of nitrogen in cereals, where nitrogen uptake and biomass increase are closely related during the course of a single growing season (see Figure 3.19). This would suggest that nitrogen uptake is essential for crop growth.

Make sure in your own mind that you understand the concepts of deficiency, and optimal and toxicity levels, as they relate to nitrogen nutrition of crops. You should, however, realise that different species can have slightly different responses to nitrogen.

3.6.5 Nitrogen cycling

In Section 3.6.1, the general principles of how soil constituents are held in, and are affected by, the soil environment are given. Nitrogen is held within a series of pools within the soil, and can be cycled between these pools. This is best described diagrammatically and the nitrogen cycle is elaborated in Figure 3.20.

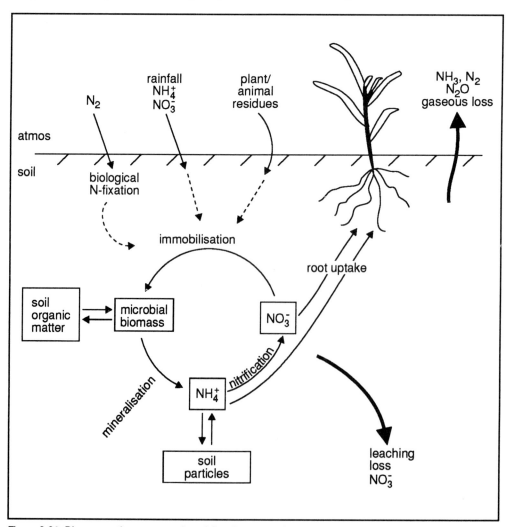

Figure 3.20 Diagrammatic representation of the nitrogen cycle.

Here we will provide an overview of the nitrogen cycle and the availability of nitrogen to plants. We will not examine the details of the biochemistry of the processes involved. Details of this aspect are dealt with in the BIOTOL text, 'Biosynthesis and the Integration of Cell Metabolism'.

As you can see, there are complex relationships between the locations of nitrogen in the environment. Nitrogen available to plants is held in 2 pools in the soil, as pools of free ammonium and as nitrate. Ammonium can be converted to nitrate by microbial action (nitrification), and nitrate can become unavailable to plants if it is taken up by a range of soil micro-organisms and converted into organic matter (humus). This process is generally called immobilisation. Humus can be converted back to ammonium by another range of micro-organisms by a process called mineralisation. This cyclical process is central to the soil phase of the nitrogen cycle. In general, more than 90% of soil N will be held as organic matter (unavailable for plant growth), and much of the ammonium will be bound to soil particles (particularly clays), also making it unavailable to plants. Consequently, the pools of nitrogen available to the plant are rather small and can easily be lost from the soil environment by volatilisation, immobilisation or by leaching (the loss of nitrogen dissolved in soil water by drainage to ground waters). The low level of available nitrogen in soils, and the ease by which it can be made unavailable, makes nitrogen the soil nutrient which most limits crop yields.

nitrification

immobilisation
mineralisation

volatilisation
and leaching

3.6.6 Nitrogen fertilisation

Agriculturalists overcome the limiting effect of low nitrogen levels by adding nitrogen to the soil. Although plant residues (and their nitrogen-containing components) in natural environments are returned to the soil, in agricultural situations this nitrogen is generally totally removed with the harvested crop. Therefore, there is a need to add nitrogen supplements to the soil. Nitrogen is normally added either as fertiliser (for example, as ammonium or nitrate salts), or by the use of nitrogen-fixing micro-organisms. These microbes are able to obtain nitrogen directly from the atmosphere.

Fertilisers added to soils to increase nitrogen levels are generally inorganic salts of NO_3^-, NH_4^+, urea, or liquid or gaseous ammonia (see Table 3.8).

Nitrogen source	% N
ammonium sulphate	21
ammonium nitrate	35
urea	46
aqueous ammonia	25-29
anhydrous ammonia	82

Table 3.8 The nitrogen content of some inorganic nitrogenous materials that may be used as nitrogen fertilisers.

The nitrogen content of these fertilisers ranges from 21-82% by weight, depending on the composition of the nitrogen compound. Organic manures also contain nitrogen, but generally at much lower levels. Typically they comprise 0.3-2.5% nitrogen (wet weight). This means that much larger quantities of organic manures, than inorganic fertilisers, have to be added to significantly raise soil nitrogen levels.

Haber-Bosch process

One of the most common forms of nitrogen fertilisers used in developed countries is ammonium sulphate ($[NH_4]_2SO_4$). This is a major product of the Haber-Bosch process in which nitrogen and hydrogen gases are reacted at high temperatures and pressures in the presence of a suitable catalyst (iron). The ammonia gas formed by this process is 'trapped' by dissolving it into sulphuric acid with the resulting formation of ammonium sulphate.

∏ The use of ammonium sulphate produced by this process as a nitrogen fertiliser has a number of economic, agricultural and environmental drawbacks. See if you can make a list of two or three of these.

There are, in fact, many practical problems which arise from the use of this fertiliser. Here we will confine ourselves to just three. The conversion of nitrogen gas to ammonia is an energy demanding process and this, of course, contributes significantly to the cost of $(NH_4)_2SO_4$ made from it. This means that this fertiliser can only be afforded by relatively prosperous farmers (ie farmers whose soil is already quite fertile). Poor farmers who cultivate soils with very low nitrogen content cannot afford such fertilisers, yet they would especially benefit from their use. We can illustrate this graphically in the following way.

Examine the graph which relates yield and nitrogen content of soil.

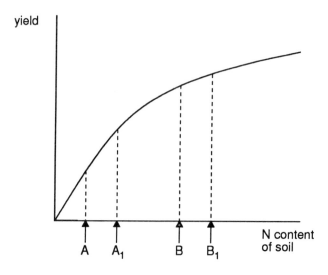

Consider a poor farmer whose soil has a nitrogen content of A. If he could afford to add ammonium sulphate to increase his soil's nitrogen content to A_1, then his yield would double. On the other hand consider a rich farmer whose soil nitrogen content is B. In this circumstance, the addition of the same amount of nitrogen fertiliser would lead to perhaps a 10% increase in yield. Thus, in global terms, farmers who will least benefit from the use of fertilisers are those who can best afford them! This same principle can be applied to many other fertilisers. This is illustrated by the data in Table 3.9 which relates global nitrogen, phosphorus and potassium fertiliser use for different regions during the 1980s. Examine this data carefully as they show the marked difference between the regions with developed market economies and the regions of so called developing market economies.

Area	N (kg ha^{-1})	P$_2$O$_5$ (kg ha^{-1})	K$_2$O (kg ha^{-1})
developed market economies			
North America	48	22	24
Western Europe	109	57	58
developing market economies			
Africa	4	4	2
Latin America	15	11	7
Near East	33	18	1
Far East	30	10	5

Table 3.9 Nitrogen, phosphorous acid and potassium fertiliser use during the 1980s. (Average values ha^{-1}). Data from Helsel, Z R 'Energy in Plant Nutrition and Pest Control' in 'Energy in World Agriculture' Ed Stout, BA, Elsevier Vol 2, 1987.

So the point we have made about fertilisers in general, and the use of ammonium sulphate in particular, is that their use, in global terms, tends not to be directed to regions where they would exert most benefit. This is not likely to alter unless a new socio-economic morality is developed within the world community.

The use of ammonium sulphate illustrates another problem associated with using mineral fertilisers. Ammonium sulphate will dissociate into its component ions in soil water:

$$(NH_4)_2SO_4 \rightleftharpoons 2NH_4^+ + SO_4^{2-}$$

During nitrification, the NH_4^+ is converted to NO_3^-. The ionic balance in the soil has to be maintained and thus protons are generated to balance the negative charges on the NO_3^- and SO_4^{2-} ions. In other words, if the soil is not buffered, there is a tendency for the soil to become more acidic.

∏ Would the addition of KCl as a potassium fertiliser lead to acidification?

The answer is yes. The plant's requirement for K is much greater than its requirement for Cl. So K$^+$ ions would be taken up and H$^+$ released (effectively producing HCl in the soil water).

We can illustrate this in the following way:

$\prod$ Would the addition of KNO_3 as a fertiliser lead to acidification?

Probably not. Plants have high requirements for both K and N (as NO_3). Thus, although potassium (K^+) ions would be taken up, so would the nitrate (NO_3^-) ions. In other words there would be no net generation of HNO_3 from KNO_3. Indeed, the soil would probably become slightly alkaline because more NO_3^- than K^+ ions are usually taken up. Therefore, OH^- ions would need to be generated to maintain an ionic balance. We illustrate this in the following way:

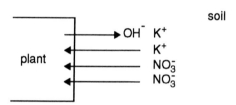

This is rather an over simplification of the events that occur in soils since we have ignored the impact of the mineral particles of soils on the ionic balance. Nevertheless, the point we have made is that when we add mineral fertilisers such as ammonium sulphate, we are not simply adding plant nutrients. The addition of these fertilisers may have many other effects which may be detrimental to crop production.

Finally in this section on nitrogen fertilisers, we need to mention the environmental consequences of using mineral fertilisers. All ammonium salts are water soluble. So are nitrates. Thus there is a strong probability that some of the added fertilisers will be lost in run off and drainage water. This not only reduces the efficiency of the fertiliser in terms of its impact on crop yield, but also causes enrichment of the receiving waters eutrophication (rivers, lakes) with plant nutrients. This process of so called, eutrophication may cause algal blooms in these receiving waters which, in turn, results in the generation of anoxia in these waters. This has important consequences on the amenity value of the water (drinking water supply, recreation value, impact on wild life).

We began Section 3.6.1 by indicating that by maintaining the levels of nutrients in soils, we should expect good crop yields. Through our discussion of ammonium sulphate as a fertiliser you should have realised that adding fertilisers improves crop yields. We will complete this section by asking the question, 'When should the fertiliser be added to the soil?'

There is no simple answer to this question. It depends both on the nature of the fertiliser and upon the crop to be grown. Slow-releasing nitrogen fertilisers (for example farmyard manure) function as slower-acting, longer-lived supplies of nitrogen than the fast-acting compounds such as ammonium sulphate or potassium nitrate.

Generally, an early nitrogen dressing stimulates leaf development and accelerates canopy closure. This is generally beneficial as it means that the photosynthetic capacity of the crop is generated early. However, care has to be taken. If, for example, nitrogen is abundantly supplied at an early stage in cereal (especially wheat) crops, this may promote stem elongation without the development of a full complement of supportive tissue. This may result in lodging (lodging = bending or snapping of the stem). This results in lower production, as CO_2 assimilation is reduced because of an unfavourable light distribution within the crop. It may also cause a greater incidence of disease and harvesting loss because of increased wetness of the leaves and air humidity surrounding the plants.

On the other hand, late additions of nitrogen fertilisers to cereals crops may lead to increased grain yields and the grain may have higher protein content.

| SAQ 3.6 | Would it make most sense to add nitrogen fertilisers to root crops early in the growth of the crop or to use nitrogen fertilisers as a late dressing? (Give reasons for your answer). |

The amount of nitrogen applied to the soil in intensive agricultural systems varies considerably, dependant upon the crop being grown, the soil type, and the previous cropping history of the soil. For a cereal crop, typically up to 400 kg ha^{-1} yr^{-1} could be added, generally as inorganic fertiliser. In extensive agricultural systems, such as tropical livestock grazing or shifting cultivations, annual nitrogen input is generally less than 20 kg ha^{-1} yr^{-1}. Where grazing occurs, nitrogen inputs tend to come from biological nitrogen fixation, and in shifting systems from the burning of forest biomass.

3.6.7 Biological nitrogen fixation

relative importance of industrial and biological nitrogen fixation

In intensive agricultural systems, addition of inorganic nitrogen as fertiliser is the standard means of raising soil levels to optimise yields. In natural ecosystems and in non-intensive agriculture, input of nitrogen from the atmosphere by biological nitrogen fixation is a much more important mechanism in quantitative terms.

Process	Annual production
industrial fixation	5 x 10^7 tonnes
biological fixation	2 x 10^8 tonnes

Table 3.10   Estimated world levels of fixed nitrogen - annual basis.

From Table 3.10 above, you can see that in quantitative terms, nitrogen fixed industrially for fertiliser addition only represents about 25% of that fixed by microbial action.

Plants themselves do not have the ability to fix atmospheric nitrogen. They rely on the ability of certain bacteria and actinomycetes; these possess the enzyme nitrogenase which is crucial to nitrogen fixation from the atmosphere. These micro-organisms are present in soils and exist either as free-living organisms or associated with plant roots. The different types of nitrogen fixing microbes are shown in Table 3.11.

System	Example of micro-organism	Energy source	Typical N yield (kg ha^{-1}yr^{-1})
symbiosis	*Rhizobium* *Anabaena*	assimilates from plant	50-600(legumes) 2-300(non-legume)
association (rhizosphere)	*Azospirillum* *Azotobacter*	root exudates	10-300
free-living	*Azotobacter*	heterotrophs - plant residues	0.1-0.5
	Klebsiella *Rhodospirillum*	autotrophs - microbial photosynthesis	up to 25

Table 3.11   N-fixing microbes and their typical yields of fixed nitrogen.

symbiotic, associative and free-living nitrogen fixers

Three different systems of biological nitrogen fixation exist. First, symbiotic relationships between bacteria or actinomycetes and plant roots exist, most forming root nodule structures on root surfaces where fixation takes place. Secondly, bacteria can live in close proximity to roots in the soil immediately surrounding the root (rhizosphere), and thirdly, bacteria can be free-living, totally independent of plants.

Biological nitrogen fixation is energy-consuming and, therefore, dependent upon a source of energy. In symbiotic relationships (the best known being the *Rhizobium* - legume symbiosis) the energy source is provided by the plant in the form of sugars; in the rhizosphere it is in the form of chemicals in root exudates (probably sugars); and in free-living systems the source is from organic matter or bacterial photosynthesis. It is likely that rates of biological nitrogen-fixation are highest where the amount of available energy is greatest. Note that the biochemistry of nitrogen fixation is described in the BIOTOL text, 'Biosynthesis and the Integration of Cell Metabolism'.

A range of environmental factors affect the rate of biological N-fixation, particularly through interacting with the activity of the nitrogenase enzyme. Nitrogenase activity is reduced when soil levels of molybdenum, calcium and phosphorus are low. It is also inhibited in the presence of oxygen, and the genes responsible for the synthesis of the nitrogenase proteins (the so-called *nif* genes) are repressed in the presence of nitrogen compounds, including fertiliser nitrogen. This means that biological N-fixing systems are not compatible with agricultural practices where large amounts of inorganic nitrogen are added to soils.

SAQ 3.7

Examine Figure 3.21. This figure plots the average yield of rice in a variety of countries in the Far East against the price ratio of brown rice and N-fertiliser. This ratio means the relative prices of rice and fertiliser.

What can be concluded from these data?

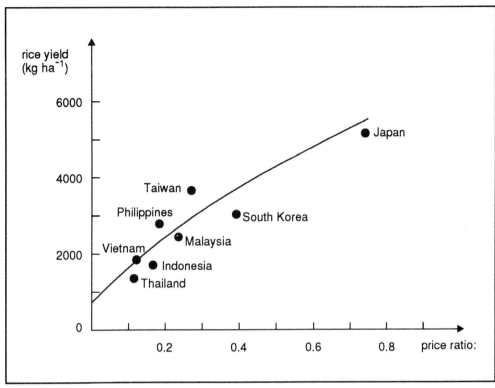

Figure 3.21   Relationship between rice yield and the price ratio of brown rice and nitrogen fertiliser in a number of countries in the 1960s.

3.7 Water and crop yields

3.7.1 The hydrological cycle

It goes without saying that sufficient water is required for the successful development of a crop to maximise its final yield. In growing plants, water can represent 75-95% of the total plant weight. Nevertheless, the role of water both in the atmospheric and soil environments is quite complicated, and is related through the hydrological cycle. The hydrological cycle summarises the main fluxes or movements of water that occur at a global level. We have provided a simplified illustration of the hydrological cycle with particular reference to plants in Figure 3.22.

In Figure 3.22, we indicated that S = the water content in the soil in the root zone. This will, of course, change as water is either added (for example by rainfall, irrigation or capillary rise) to the system, or removed (for example, by drainage).

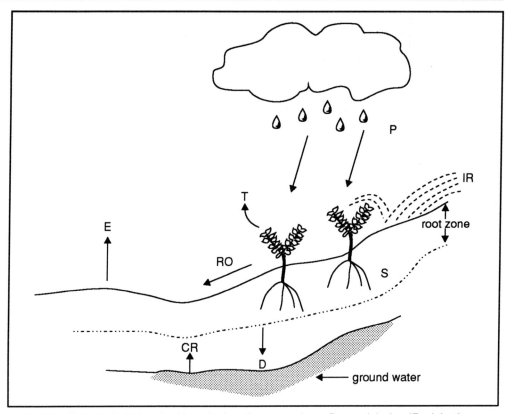

Figure 3.22 The hydrological cycle with particular reference to plants. P = precipitation; IR = irrigation;
RO = lateral run off (or run on if flowing from an adjacent area); T = transpiration; E = evaporation from the
soil surface; D = drainage from the root zone; CR = vertical upward flow from soil layers below the root
zone by capillary action (capillary rise); S = the water content in the root zone.

∏ See if you can write an equation relating the factors described in Figure 3.22 to a
change in S.

The equation we hope that you deduced is:

$$\Delta S = P + IR + CR \pm RO - T - E - D$$

$$\underbrace{}_{\substack{\text{adding water} \\ \text{to the system}}} \quad \Big| \quad \underbrace{}_{\substack{\text{removing water} \\ \text{from the system}}}$$

$$\pm \text{ RO}$$

(depends on whether
this represents run
off or run on)

∏ Suggest suitable units for the terms used above.

Suitable units would be kg H_2O m^{-2} day^{-1} (that is, amount of water per unit area of soil
surface per unit time). These are not the only units you could have used, for example,
mm H_2O day^{-1} could also be used. (Note mm = volume ÷ area).

If, for the moment, we ignore irrigation, the features of the hydrological cycle which have the greatest influence on crop production at any given locality are:

- total rainfall per year;

- water requirements of the crop;

- water losses to the atmosphere by evaporation and transpiration.

∏ How do total rainfall, crop water requirement and water losses interact to determine whether crop production for a particular crop species is economic at a given location?

seasonal variations in the availability of water

When deciding whether a given location will be suitable for production of a particular crop, one has to consider not only the total amount of yearly rainfall, but also at what time of year the rain falls, and its reliability. For example, in Mediterranean regions, rain falls mainly in the winter. Crops grown in these regions have to be able to grow in the colder winter months, or must be able to use the relatively low levels of water stored in the soil for growth during the hot, dry summers. In this situation, fast growing summer annuals, such as cereals, have become favoured by farmers because they are able to efficiently use soil water to complete their growth cycles, ie, to come to harvest, prior to the intense heat of high summer. Where water does limit crop growth, such conditions

drought

are normally termed drought.

Different crops have differing water requirements which change as the crop develops. For example, maize requires between 30 and 100 mm of water per month during its growing season, with the greatest requirement when the crop is growing fastest and in the hottest months. It must be remembered that if insufficient water is available, leaf stomata will close which will cause photosynthesis to cease. The assimilation of carbon via photosynthesis is central to plant growth and the maximisation of crop yield. In

water use efficiency

drought conditions, those crops with a more efficient use of water tend to give higher yields as they generally can keep stomata open longer under dry conditions.

The effectiveness in moderating water loss while allowing sufficient CO_2 uptake for photosynthesis can be assessed by determining the transpiration ratio:

$$\text{transpiration ratio} = \frac{\text{amount of water transpired}}{\text{amount of } CO_2 \text{ fixed}}$$

The reciprocal of transpiration ratio is often used and is called the water use efficiency (WUE).

SAQ 3.8

For a wheat leaf transpiring at a rate of 50 mg H_2O m^{-2} s^{-1} and photosynthesising at 0.6 mg CO_2 m^{-2} s^{-1}, what is the water use efficiency of that leaf?

(Express this in terms of mass of water used to assimilate 1 g of CO_2).

The units used for water use efficiency (WUE) are variable. WUE can be expressed as the ratio of the mass of CO_2 fixed per unit area per unit time to the mass of water transpired per unit area per unit time (see SAQ 3.8). Alternatively it may be expressed in terms of mass of economic product per mm of transpired water. Table 3.12 lists a number of WUE values for a range of crops.

Crop	WUE	Comments
wheat	10-25	C3 assimilation
potato	15-25	C3 assimilation
maize	20-40	C4 assimilation
sorghum	25-50	C4 assimilation

Table 3.12 Value of WUE (in kg of economic product/mm transpired H_2O/per ha) for a variety of crops. (Note that the value of transpired water is given in mm. It is calculated from volume of transpired water ÷ ground area of crop).

3.7.2 Availability of water and soil type

In Section 3.7.2, we described the hydrological cycle in general and explained that crop production is mainly influenced by rainfall, loss of water by evaporation and transpiration, and the efficiency by which plants may use water (the WUE).

The amount of water held in the soil is also dependent on the type of soil. Water is held in the soil in three forms:

- **gravitational water** - available to plants but quickly draining away. Generally present only after rain;

- **capillary water** - water held in capillary pores between soil particles held by surface tension. Available to plants;

- **hygroscopic water** - bound tightly to soil particles and not available to plants.

After profuse rain, water is present in soil in all three forms. Obviously, water will drain away with time so that only water bound or held by surface tension will be present. This point is called the field capacity of the soil. Gradually, evaporation and transpiration will remove the capillary water from the soil, so that the amount of water available to crops becomes less and less, to a point where only hygroscopic water is present. This is known as the permanent wilting point because at this soil moisture level most plants are not able to take up water and hence wilt (a sign of plant water stress). Soil type alters the precise points of field capacity and permanent wilting point, as shown in Figure 3.23.

field capacity

and

permanent
wilting point

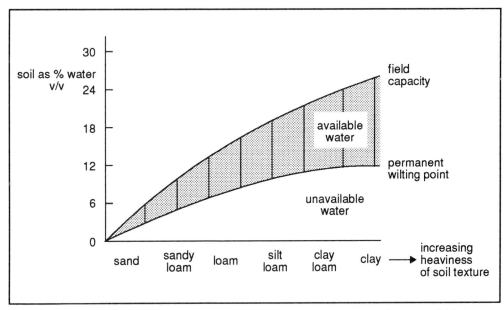

Figure 3.23 The influence of soil type on field capacity, permanent wilting point and water available for plants to use (see text for discussion).

Let us consider the water availability in soils in a little more detail.

Sandy soils are characterised by a large fraction of course particles (50 µm) which means that their pores (regions between particles) are large. Water, therefore, drains quickly from such soils. Clay soils, on the other hand, have much smaller particles (2 µm) and the pores between them are much smaller. The narrow pores form a high resistance to water transport. Thus drainage losses, shown as D in Figure 3.22, are small. The large surface areas of the particles in clay soils coupled with the fine pores, promote capillary rise (CR in Figure 3.22). However precipitation or irrigation infiltrates only slowly. Thus, on slopes, clay soils often show higher run-off losses.

SAQ 3.9

A clay soil has a field capacity of 25% (v/v) soil water. The extent of precipitation during the growing season maintains the water content at the field capacity throughout the growing season. Is this soil likely to support the production of a high-yield crop? (Give reason(s) for your answer).

3.7.3 Water availability and plant responses

Plants also differ in their responses to water, depending on the water status of the environment in which a species is adapted to live. Three groupings are generally recognised:

* **hydrophytes** - plants of waterlogged environments;

* **mesophytes** - plants of normal rainfall regimes;

* **xerophytes** - plants adapted to regular drought.

Nearly all crop plants are mesophytes, with no important agricultural species showing classical xerophytic characteristics. Xerophytic characteristics generally include one or more or the following features: sunken stomata, protective leaf hairs, rolled leaves, large amounts of root in comparison to shoot, reduction in leaf volume, succulence. In general, xerophytic adaptations lead to a reduction of water loss and conservation of stored water. Nevertheless, a range of drought tolerance is shown, with some crops being tolerant to some extent of low water availability. The crops grown where water availability is low are generally quick growing to make use of any periodic or seasonal rain fall. Such cultivars tend to be genetically lower-yielding than cultivars adapted to normal rainfall, and may also show yield reduction due to lack of water itself.

The physiological effects of drought on mesophyte crop species is widespread and complicated. The range of plant processes affected by low water availability is shown in Table 3.13.

Process or parameter affected	Sensitivity to stress			Remarks
	very sensitive ←————— —————→ relatively sensitive			
	Tissue ψ required to affect process			
	0 MPa	-10 MPa	-20 MPa	
cell growth	——— - -			fast-growing tissue
wall synthesis	———			fast-growing tissue
protein synthesis	———			elongated leaves
protochlorophyll formation	——			
nitrate reductase level	———			
abscisic acid accumulation	- - ———			
cytokinin accumulation	——			
stomatal opening	- - ——————————— - - -			depends on species
CO$_2$ assimilation	- - ——————————— - - -			depends on species
respiration	- - ———			
proline accumulation	- - ———			
sugar accumulation	———			

Table 3.13 The sensitivity of plant processes to water stress. ψ = osmotic pressure. Length of the horizontal lines represents the range of stress levels within which a process first becomes affected - dashed lines signify deductions based on more tenuous data. (Adapted from Salisbury and Ross, Plant Physiology, Wadsworth, Belmont Ca, 1992).

The amount of water in the soil available to the crop depends not only on the total rainfall, but also on the extent to which water is lost from the soil back to the atmosphere. Water loss from the soil is due to evaporation from the soil surface and transpiration, and together is termed evapo-transpiration. Drought will occur when evapo-transpiration exceeds input from rain, hence leading to yield reduction.

∏   Do you think there is any way that a farmer can overcome a situation where water availability is limiting?

Where insufficient water is available to the farmer to grow a given crop, two options are open to him. Firstly, a crop or a variety could be selected that would perform better under the given conditions. This might only be possible where water is only mildly limiting. Additionally, alternative crops may not be satisfactory for economic or cultural reasons. For example, wheat does not yield well under drought conditions, but is still grown by subsistence farmers in areas subject to drought because it is the only cereal that is suitable for bread making.

The second option is to correct the water shortfall through irrigation. In the developing world, water is often added to fields in the flood plains of rivers such as the Nile or the Indus, generally via irrigation canals or ditches. In more modern systems, water is pumped in pipes and distributed by sprinkler or trickle systems. The design and operation of irrigation systems is beyond the scope of this chapter, except to point out that substantial increases in yield are possible by using irrigation.

Summary and objectives

In this chapter we have examined the influence of external factors on crop productivity with particular emphasis on the physical and chemical environment. We described how climate is the major factor which governs which crops may be grown. We also explained how the leaf canopy may itself affect the microclimate of a crop. We provided a detailed discussion of the influence of solar radiation on crop yield and we examined the effects of nutrient and water availability on crop growth. The practices of mineral fertilisation were discussed.

Now that you have completed this chapter you should be able to:

- distinguish between plant and environmental components which influence crop yield;

- list the main types of environmental factors which determine crop yield;

- distinguish between climate and weather;

- explain how the interception of solar radiation is influenced by the crop canopy and the morphology of plants;

- use and interpret data relating to incident radiation and CO_2 assimilation;

- list the essential plant nutrients and distinguish between macro- and micronutrients;

- explain the role of nitrogen in plant nutrition and explain why nitrogen availability is often yield-limiting;

- give examples of nitrogen fertilisers and explain some of the problems which might arise from their use;

- list factors which influence the availability of water within soil;

- explain why adequate (but not excessive) supplies of water are needed to maximise crop production.

Factors affecting crop productivity: the biotic environment

Factors affecting crop productivity: the biotic environment

4.1 Introduction

Crop production and yield are affected by a range of environmental conditions. Maximal production occurs when a range of environmental factors such as temperature, rainfall and day length, are at optimal and balanced levels for a given crop species. The effects of these factors have been described in the previous chapter, but it is important to note that these factors are all of a physical or chemical nature.

In most agricultural systems, maximal (or potential) crop yield is rarely, if ever, reached. This is frequently due to physico-chemical (abiotic) factors being sub-optimal or stressful for a given crop genotype. Nevertheless, these factors are not the only causes of yield reductions. In a great many agricultural environments, especially when growing conditions are favourable, biological (biotic) factors are major causes of yield loss.

Worldwide, losses in food production from the effects of biological factors on crop yield are considerable, with perhaps up to 35% of all production being lost through disease, pests and weeds. These losses can occur both during the growth and harvesting of a crop, and during the storage of the food material prior to consumption. Up to 20% of production can be lost during storage, this being a particular problem in tropical countries.

The pie chart shown in Figure 4.1 gives you some estimate of total losses based on data from the UN's Food and Agriculture Organisation. If you were to add yield losses from other environmental factors, then perhaps only 30-40% of potential world food production reaches the consumer.

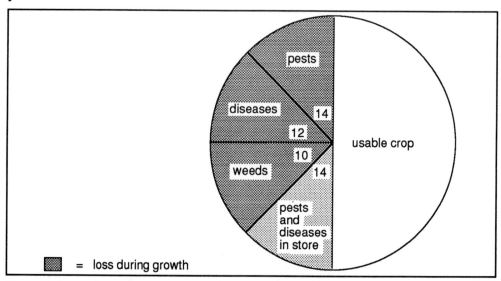

Figure 4.1 Crop losses during growth and storage. Figures are % loss.

Π To make sure that you understand the difference between biotic and abiotic factors affecting crop yield, identify which of the following factors are biotic and which are abiotic:

wind; plant diseases; light; plant competition; insect pests.

You should have identified plant diseases, plant competition and insect pests as being biotic. In this chapter, we will discuss the loss of crops arising from biological factors. We will first examine losses arising from diseases. You should realise that this is a very large topic. Here we will describe the general features of crop diseases but will also include some discussion of the range of disease-causing organisms and the effects that they have on their hosts.

We will then go on to examine the effects of pests on crops. We will briefly outline the range (invertebrates and vertebrates) of organisms that may be regarded as pests before examining the economic consequences of pest damage. We will also examine the strategies that are used to reduce the losses arising from pest damage. In the final part of the chapter we will discuss the consequences of competition between plants, especially the reduction in crop yield arising from competition from weeds.

This is a long chapter so do not attempt to study it all in one sitting.

4.2 Biotic factors causing yield reduction

The range of biological factors that can affect crop yield is wide, but broadly falls into three groupings:

- plant diseases;

- plant pests;

- plant competition.

Π See if you can write down named examples of each.

(Do not worry if you can not do this with any real confidence at this stage. By the end of this chapter you will be able to cite many examples).

plant diseases mainly caused by fungi, bacteria and viruses

Plant disease is generally described as an infection of a plant species by a pathogenic (disease-causing) organism. The most important organisms causing disease are fungi, bacteria and viruses, although there are a range of parasitic higher plants such as *Cuscuta*, *Striga* and mistletoe which can be considered as diseases through their ability to reduce yield. The symptoms of mineral deficiencies or salinity may sometimes be considered to be evidence of a disease. However, from their non-biological nature, these are best considered as 'physiological disorders' and are obviously caused by an imbalance of some physico-chemical factor. As such, we will not consider them here as 'diseases'.

invertebrate and vertebrate plant 'eaters'

Plant pests are considered broadly within two areas, smaller plant consumers, such as insects, molluscs and nematodes; and larger, principally vertebrate, plant consumers. The effect of pests is either to consume useful plant yield, or to reduce the ability of a crop to maximise yield by consumption of the root and/or shoot capacity, which would subsequently contribute to the formation of the final crop yield. Also, some pests may have little effect themselves on the physiological capacity of the plant, but serve to transfer disease between plants, thus having an indirect effect on yield.

Although most invertebrate pests can be considered as 'herbivores', this term is usually used to refer to vertebrate consumers of plant material. Many agricultural systems rely on the use of animals to convert plant material into bodyweight for human consumption, so that the management of animal grazing to maximise pasture production is important; pasture production can be reduced by grazing of non-useful species.

competition between plants

In the case of competition, plants themselves can act to alter the yield of a crop. All plants have requirements for a range of physical factors for optimal growth, such as light, water and nutrients, so that when plants are growing as a crop, ie within close proximity to other plants, competition will occur for the finite levels of resource available from the external environment.

Two types of plant competition are recognised:

• intraspecific competition;

• interspecific competition.

Intraspecific competition represents plants of the same species competing for resources within a crop stand, whereas interspecific competition is competition by different species for the same resources.

4.3 The effect of plant disease on crops.

Definition

Plant disease can be defined as a significant deviation from the normal physiological processes of plant function, due to the presence of an agent which causes disease (the causative agent).

In terms of crop production, the presence of disease can lead to a quantitative reduction in yield, or a reduction in crop quality. Although this definition could include a 'physiological disorder', it really refers to reductions caused by a biological agent, that is, a plant pathogen.

Table 4.1 shows the main groups of organisms which cause plant diseases.

You can see that a large range of disease-causing agents exists. The examples given in the table represent a small fraction of the total number of economically important disease-causing organisms, which runs into many thousands. This range of disease-causing organisms has varied effects on crop production, affecting a range of physiological processes leading to yield reduction, as is shown diagrammatically in Figure 4.2.

∏   From your knowledge of plant physiology you should be able to explain, at least in outline, why diseases which affect particular physiological processes, may produce the kinds of effects that are described in Figure 4.2. For example, viruses which infect leaf tissues are likely to damage the cells containing the photosynthetic machinery. Thus infected leaves often appear yellowed or mottled in appearance. See if you can explain how the other effects shown in Figure 4.2 may arise. You will have an opportunity to check your ideas later in this chapter.

Group	Sub-group	Typical diseases
viruses		barley yellow dwarf virus (BYDV) virus yellows in sugar beet tomato mosaic virus (TMV)
bacteria		bacterial cankers soft rots black rot diseases
fungi	myxomycetes	club root in Brassicas potato wart disease
	phycomycetes	diseases caused by species of *Peronospora* (eg, downy mildews), *Phythium* (eg, watery wound rot in potato), or *Phytophthora* (eg blight in potato)
	ascomycetes	various powdery mildews *Septoria* diseases *Helminthosporium* diseases (eg leaf blight of maize)
	basidiomycetes	rusts, smuts and bunts in cereals

Table 4.1 Major groups of organisms which cause diseases in plants.

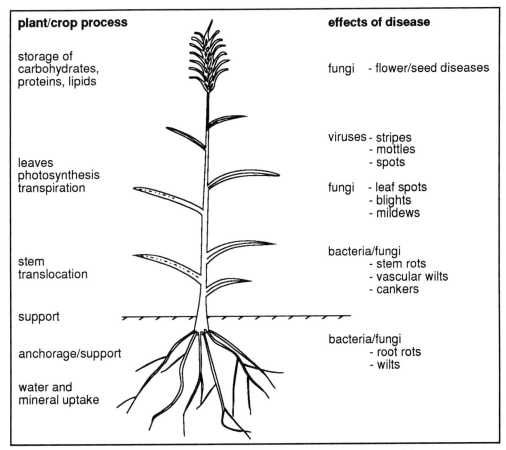

Figure 4.2 Diagrammatic representation of some physiological processes in plants and the effects of disease. We have given an indication of the main groups of disease organisms which cause the effects listed on the right-hand side.

4.3.1 The pathogen life cycle

During the course of disease development, pathogens generally go through a number of stages of development, which is normally called the life cycle of the disease-causing organism. Such life cycles have generally evolved to ensure that the pathogen is able to infect subsequent generations of the plant host, and consequently these life cycles differ in complexity dependent upon the nature of the interaction of the pathogen with the host. Such life cycles can vary from a purely vegetative state with asexual reproductive stages, to highly complex life cycles where asexual and sexual reproductive phases take place in different host species.

late blight disease of celery

An example of a simple life cycle is that of the late blight disease of celery, caused by the Deuteromycete (Fungi Imperfecti) *Septoria apiicola*, which is one of main yield-reducing pathogens of the celery crop (Figure 4.3).

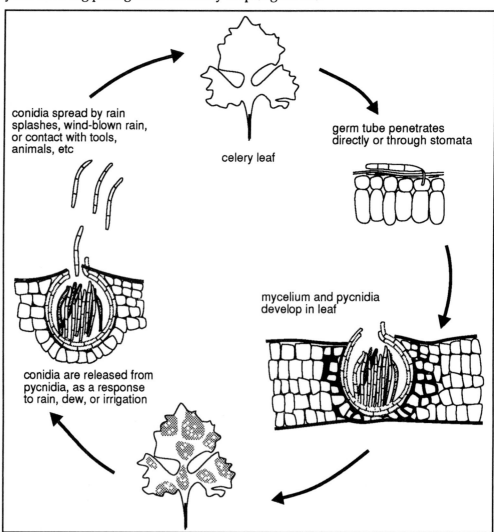

conidia spread by rain splashes, wind-blown rain, or contact with tools, animals, etc

celery leaf

germ tube penetrates directly or through stomata

mycelium and pycnidia develop in leaf

conidia are released from pycnidia, as a response to rain, dew, or irrigation

Figure 4.3 The life cycle of the late blight disease-causing organism, *Septoria apiicola*, of celery.

black rust disease of cereals

This is, comparatively, a very simple life cycle. In contrast, the causal organism of the black rust disease of cereals (*Puccinia graminis*), which is a Basidiomycete, has a highly complex life cycle (Figure 4.4).

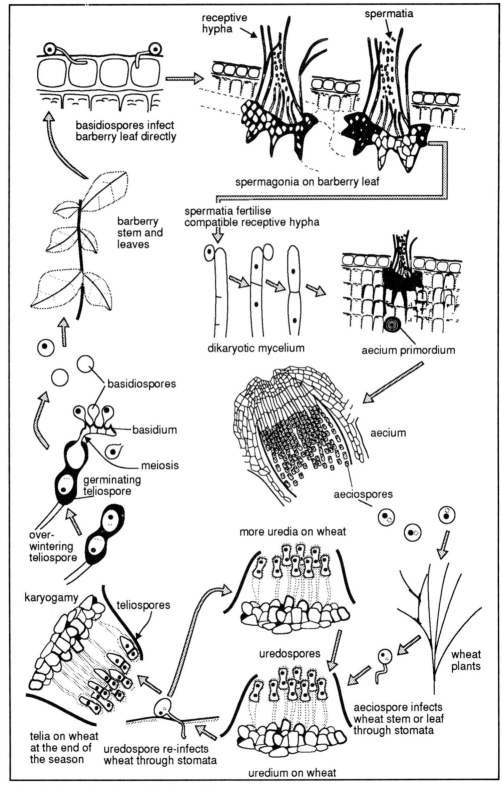

Figure 4.4 The life cycle of *Puccinia graminis*.

Although a detailed knowledge of individual life cycles is not necessary in the context of this book, do note the complexity of the last example. There are 4 spore types, asexual and sexual reproductive phases via these spores, and two host species.

basidia
basidiospore

Let us follow the life cycle in a little more detail. We will begin with the basidiospores which are released in Spring from basidia which develop from over-wintered teliospores (left-hand side of Figure 4.4). Basidiospores which alight on barberry stems and leaves germinate and produce hyphae. These hyphae differentiate to produce

spermagonia
spermatia
aecia
aeciospores

spermagonia which contain the so-called receptive hyphae and the spermatia-forming hyphae. When spermatia are produced, they combine with compatible receptive hyphae which develop into aecia which produce aeciospores. Aeciospores are wind blown, and if they alight on a wheat plant, they germinate to produce hyphae which ramify the stem and leaf tissues. Note that this infection enters the wheat via stomata.

uredia
urediospores

In the wheat-tissue, the fungus grows rapidly and begins to produce special fruiting bodies (uredia) in which urediospores are produced asexually. Urediospores infect available wheat plants. These types of spore, therefore, represent a rapid propagation system for the fungus. The process of uredospore production, release and re-infection continues whilst weather conditions remain favourable and there are still wheat plants to infect. Thus rapid propagation may lead to severe crop damage and the whole crop may be covered in the black uredia.

telia and
teliospores

Towards the end of the growing season, another type of fruiting body is produced, the teleium. Telia produce teliospores which over-winter on the wheat.

∏ The life cycle shown in Figure 4.4 looks quite complicated because of the various fruiting body and spore sturctures have been shown. It might help you to remember the order of spore formation by completing the flow diagram of the life cycle of *Puccinia graminis* drawn below.

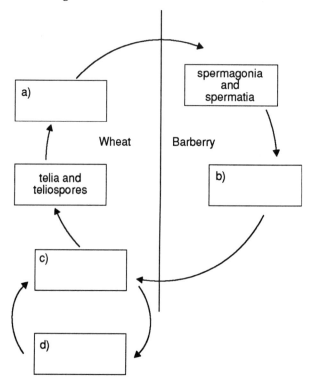

You should have used the following labels:

- a) basidia and basidiospores;

- b) aecia and aeciospores;

- c) and d) uredia and urediospores.

Many disease organisms, especially fungi, show similarly complex life cycles. It is important to agriculturists that the details of such life cycles are known, since this may enable them to develop strategies to control the disease. Let us use *Puccinia graminis*, shown in Figure 4.4, as an example.

Π See if you can suggest two ways of reducing the damage caused to wheat crops by *Puccinia graminis* without using a fungicide or using *Puccinia*-resistant wheat strains.

There are several options here, but the two that are most likely to be successful are:

- remove barberry plants from wheat-growing areas. Without this second host, the life cycle of *Puccinia graminis* will be broken and it will fail to propagate;

- remove and destroy (burn) straw from wheat fields at the end of the season. In this way telia will be removed from the fields and, consequently, no basidiospores will be produced in the following spring.

We see in this single example how knowledge of the life cycle of plant pathogens is often of vital importance in controlling the extent of a plant disease. We will return to this aspect of plant diseases later.

4.3.2 Disease symptoms

signs of disease = symptoms

The infection of a plant by a pathogen leads to changes in the physiology of the plant and these changes give rise to the symptoms (signs) of the disease. The effect of the disease on crop yield will depend upon the extent of the infection and the severity of the disturbance of the normal physiology of the plants. Thus diseases which cause only minor disturbance to normal physiological functions will most likely have only a minor effect on yield whilst infection which greatly disrupt normal function will have a major impact on yields.

The nature and extent of physiological disturbance and hence the symptoms of the disease will depend upon the invading pathogen and the host plant species. The extent of the expression of disease symptoms varies considerably depending on a range of factors including:

- the natural resistance of the host;

- the host's nutritional history;

- the host's developmental stage.

- the physical environment.

This has led plant pathologists to develop the concept of the disease triangle, where the properties of the pathogen, the host and the environment interact to determine the course of disease development.

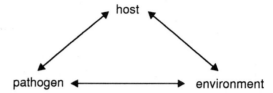

The balance between these factors can determine whether or not a potential outbreak of disease turns into a serious infection which could lead to a reduction in crop yield.

We can illustrate the disease triangle using the example of the late blight disease of potato, a potentially devastating disease that has led to potato famines throughout the world (see Figure 4.5).

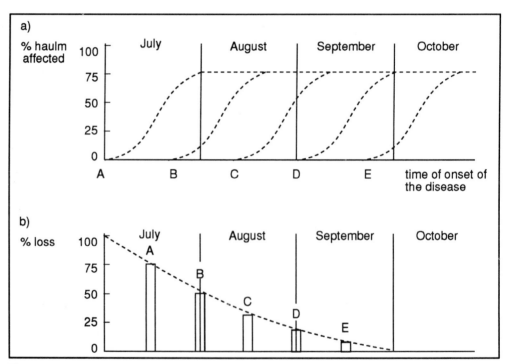

Figure 4.5 Diagrammatic representation of the loss in yield caused by late blight disease of potatoes in relation to the time of onset of the disease and the percentage of haulms affected. a) Spread of disease symptoms through haulms show the same pattern at whatever time of year infection occurs but (b) the later the infection the lower the loss.

Π In Figure 4.5 we showed that early outbreaks of a disease cause the greatest loss in crop production. Given the information, what is critical to reducing crop loss by such infections?

disease forecasting

The key point is to ensure the crop does not become infected early in the season. Obviously, it would be most useful if farmers and growers could be warned of a likely disease outbreak. To this end, the science of disease forecasting has developed, with the object of predicting disease outbreaks so that the preventive control measures can be applied. Details of forecasting procedures generally involve computer modelling of disease epidemics using current disease and environmental data. 'High risk' periods can be identified and growers warned.

Disease symptoms can take many forms and, of course, are characteristic of a particular disease, and will also depend upon the host species. Most pathogens will cause one or more symptoms, and for the diagnosis of that disease, it is important to assess the composite pattern of symptoms to ensure correct diagnosis. Remember that incorrect diagnosis could lead to inappropriate control measures being applied.

Common symptoms include:

- **chlorosis** - localised or general yellowing of leaves due to chlorophyll degradation;

- **necrosis** - cell death. Depends on the location in the plant and can lead to leaf spots and lesions, blights, scabs and rots;

- **changes in growth and development** - common effects of cankers and viruses, may be localised or general;

- **wilting** - caused by root damage, or blockage of water transport in the vascular tissues.

4.3.3 Invasion of the host by the pathogen

Most fungal pathogens are spread from plant-to-plant within the crop in the form of spores. Pathogens produce a huge range of spore types dependent upon species, some asexual (eg conidia) and some as part of the fungus's sexual stages (eg basidiospores). For our purposes, we can just consider spores as infective agents central to the transmission and spread of fungal plant diseases. For an invasion to be successful, the pathogen has to pass through several stages before complete colonisation of the plant can take place and spores can be released to cause further plant infection. The three main stages are: spore germination, penetration of the host and colonisation. We will examin............e each in turn.

transmission by spores

Spore germination

Spores have to successfully germinate at the site of contact with the host. This is often dependent on physical and chemical factors such as the presence of water, and the availability of nutrient sources such as sugars on the leaf surface and root exudates in the rhizosphere.

Penetration of the host

After germination, the pathogen has to enter the tissue system of the host (ie penetration). Three routes of entry are commonly used by pathogens to enter the host:

- entry after wounding of the host - many pathogens do not actually have the ability to directly penetrate the host and are reliant upon some other 'wounding' agent. Wounding agents can include damage caused by machinery during cultivation, pest damage, damage from the physical environment (eg wind, rain, hail, etc) or from prior pathogen injury;

- entry via natural openings on the plant - plants possess a variety of direct openings from the external environment through which pathogens can enter directly. These include stomata, lenticels and nectaries. This a favoured method of entry by bacterial pathogens;

- penetration directly through the cuticle - the cuticle is a complex structure made up of layers comprising assortments of waxes, polysaccharides and cutin. A typical plant cuticle is shown in Figure 4.6.

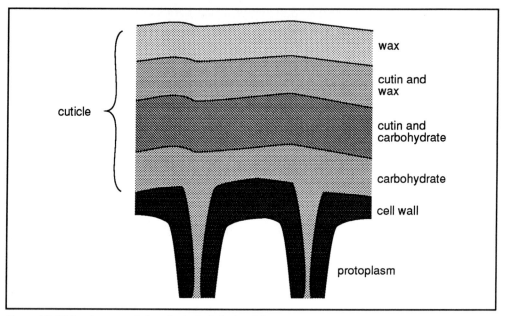

Figure 4.6 Stylised representation of a typical plant cuticle.

Π Examine Figure 4.6 carefully and see if you can suggest how a fungal pathogen might gain access to a plant through the cuticle.

appressorium For fungi to penetrate this structure, the germ tube or hypha from the germinated spore has to both exert a considerable force as well as employ a whole range of enzymes to degrade the cuticle. In most cases, the fungal hypha attaches itself to the cuticle by a specialised swollen hypha called the appressorium. From the lower surface of the appressorium, an infection peg is produced, which physically penetrates the cuticle, with the aid of softening enzymatic secretions.

cutinases
cellulases
pectinases In order to degrade cutin within the cuticle, an enzyme called cutinase is secreted. Once through the cuticle, the fungus encounters the cell wall, which largely contains polysaccharides (celluloses, hemicelluloses and pectins), as well as lignins (phenolic compounds) when it is secondarily thickened. To degrade cell walls, pathogenic fungi produce cell wall degrading enzymes, particularly a range of cellulases and pectinases.

Colonisation of the host by the pathogen

After penetration, the pathogen is able to enter and extend through the host plant's internal tissues. The amount of damage caused is dependent upon:

- the pathogen species - some pathogens, for example powdery mildews, are largely surface colonisers, not extensively ramifying through the leaf tissue. At the other extreme, some pathogens such as the vascular wilts are systemic and can ramify throughout the plant;

- environmental conditions - adverse environmental conditions can restrict pathogen development within a host;

- host resistance mechanisms - if a host has a degree of resistance to a particular pathogen, the spread of that pathogen may be reduced or halted. Host resistance to pathogen attack is complex, and it is considered in more detail in a later section.

4.3.4 Effects of pathogens on the normal physiological processes of a plant

changes in metabolism

Before we explore how host plants are able to resist pathogen attack, we must first examine how pathogens affect their hosts in more detail. In diseased plants, a whole range of critical physiological processes are altered. In general terms, after infection has become established, there is normally a decrease in CO_2 fixation (photosynthesis), an increase in respiration, and changes in the regulation of starch, amino acid, protein and phenol metabolism.

sub-threshold and sub-lethal infection

In many plant-pathogen interactions, these changes to metabolism are not lethal (to the plant). There is a threshold level, applicable to most crops, below which infection will not produce a detectable loss in yield. Above this level, although an infection may be sub-lethal, the level of infection is such that it may debilitate the host to such an extent that yield is reduced. In other interactions or when the infection reaches an advanced stage, one or more of the critical biochemical pathways fail(s) and cell/tissue death follows. Tissue death may quickly become organ death such as the loss of leaves (for example, foliar pathogens) or the destruction of storage tissue (for example, rhizomes or root crops). These all lead to a loss in biological and economic yield and can also reduce the quality of the crop product.

'damping off' disease

Reductions in photosynthetic area (for example in the case of foliar pathogens, such as late blight on celery, which cause extensive loss of green photosynthetic tissue on aerial parts) lower the capacity of plants to fulfil their potential in terms of yield. 'Damping off' diseases (which cause the death of seedlings in large numbers or of individual plants in a crop stand) reduce yield proportionately to the number of individual plants lost, although there may be some compensation from increased growth as plants have more space (as intraspecific competition is reduced in the crop stand, see Section 4.5.1). Yield may also be directly affected by pathogens which transform, replace or totally destroy the harvestable product (for example, the sclerotia of the ergot fungus *Claviceps purpurea* completely replaces the seed grain in the inflorescence of cereals and thus reduces the grain yield).

In general, infections that are capable of causing changes to the normal physiology, anatomy and morphology of the host plant (whether singly or in combination) are likely to cause effects leading to yield depression - both in terms of biological yield and economic yield.

The effect that pathogens can have on susceptible plants may be conveniently divided into:

- effects on respiration;

- effects on photosynthesis;

- effects on translocation;

- effects on growth.

We will discuss each in turn.

Effects on respiration

An increase in plant respiration has been reported for many host:pathogen interactions involving fungi, bacteria and viruses, although most information has come from studies involving biotrophic fungi.

There are, for simplicity, two types of host:fungal interaction:

- biotrophic;

- necrotrophic.

biotrophs and necrotrophs
We may define biotrophs and necrotrophs in the following ways. Biotrophs - parasitic organisms which obtain their nutrient supply only from living host tissue. Necrotrophs - parasites that cause the immediate death of host cells as they pass through them but are able to colonise dead tissue.

saprophytes
A third group of parasites are saprophytes which use organic matter from dead tissue as a food source but do not kill the tissue themselves.

In plants infected with biotrophic pathogens, respiration rates can double. In catabolism, larger organic compounds (for example starch and sugars) are oxidised to simple compounds, releasing energy that drives all other physiological processes. In the first phase of catabolism, glucose is converted to pyruvate by glycolysis or by the pentose phosphate pathway. The second phase is the Krebs (TCA) cycle. The reduced pyridine nucleotides (especially NADH) generated by this process are re-oxidised by molecular oxygen via an electron transport chain. This oxidation process is coupled to the phosphorylation of ADP to form ATP in a process usually referred to as oxidative phosphorylation. The overall effect of respiration is to generate energy as ATP to drive energy-requiring processes within the plant.

In healthy plant tissues, the glycolytic pathway and Krebs cycle predominate. Although changes in both glycolysis and Krebs cycles have been observed in diseased plants, it seems that there is a shift in respiratory balance from glycolysis to the pentose phosphate pathway. Several possible mechanisms have been proposed. These include:

- an uncoupling of oxidative phosphorylation.

The supply of adenosine diphosphate (ADP) is crucial to the regulation of cell respiration in plants. In healthy (non-infected) plants, the level of ADP is kept low by the conversion of ADP to adenosine triphosphate (ATP) by the operation of the electron transport chain. Increases in respiration can be induced artificially in plants when the formation of ATP is uncoupled from electron flow. This is most commonly done by treating plants with 2,4-dinitrophenol (DNP). DNP dissipates the pH gradient across the respiratory membranes. Thus the respiratory electron transport chain is no longer acting against a pH gradient and it speeds up. As there is no longer a pH gradient across these membranes, then there is no proton motive force to drive ATP synthesis, so ATP levels fall.

Infected tissue is often characterised by a low level of ATP and a high rate of respiration and it is thought that the pathogen mimics, in some way, the action of DNP. This has been shown in barley infected with the net blotch pathogen.

In biotrophic interactions, the increase in the observed respiration rates is due to the pathogen's own high requirement for ATP. The pathogen absorbs metabolites from the cells of the host and uses them to generate its own ATP.

In contrast, many necrotrophs produce toxins. The fungally produced toxin, victorin, when applied to plant tissue, has been shown to cause a rapid loss of ions and a subsequent increase in respiration.

- Changes in the relative importance between glycolysis and the pentose phosphate pathway.

  Evidence exists for the operation of the non-glycolytic pathway. After infection, although there may be a small increase in the activity of glycolytic enzymes, there is often a 2-3 fold increase in the activity of enzymes of the pentose phosphate pathway.

- Stimulation of the host's metabolism.

  Infection causes a general stimulation in host metabolism. Increased demand for nutrients stimulates a synthetic metabolism which in turn stimulates respiration. This can lead to extra growth, which can lead to a degree of compensation for infection in terms of crop yield.

Effects on photosynthesis

The most common effect of infection on aerial parts of a plant is a reduction in the area of photosynthetic tissue. This adversely affects growth and the biological yield and economic yield are potentially reduced. With non-foliar pathogens, there may be effects on photosynthesis that do not appear as visible chlorosis/necrosis (ie with no apparent loss in photosynthetic tissue).

Although photosynthesis is now well understood, the means by which it is controlled by the plant and modified by the pathogen are not. Very often a reduction in photosynthetic tissue does not always translate into a reduction in photosynthesis. For example, broad beans have been known to suffer 30-40% leaf loss due to pathogenic attack while maintaining relative growth rates similar to those of healthy plants. Furthermore, areas of leaves adjacent to infected tissue have been shown to increase their photosynthetic efficiency to compensate for the losses arising from the infected tissue.

importance of chlorophyllase

Early workers concentrated on the breakdown of chlorophyll by the enzyme chlorophyllase which is often followed by the breakdown of chloroplasts, resulting in chlorosis.

$$\text{chlorophyll} \xrightarrow{\text{chlorophyllase}} \text{chlorophyllide} + \text{phytol}$$

In chlorosis associated with viral infections, higher levels of chlorophyllase have been detected. Most viral infections that induce chlorosis, reduce not only the total amount of chlorophyll but also the efficiency of the remaining chlorophyll. This, of course will affect photosynthetic rate and hence yield potential.

In fungal and bacterial infections, loss of chlorophyll occurs. However, the remaining chlorophyll remains fully active. In some cases, especially in plants infected by biotrophs, tissues immediately adjacent to the fungus have increased photosynthetic activity while the rest of the leaf turns chlorotic. These are often termed 'green islands'. They may be due to selective retention of chlorophyll in areas around the pathogen or it may be that the chlorophyll is resynthesised more rapidly than it is broken down. The retention of chlorophyll is thought to be due to the secretion of cytokinins by the biotroph.

green islands

A distinction between necrotrophs and biotrophs is that the net assimilation rate is reduced rapidly in necrotrophic infections whereas in biotrophic interactions there may be an increase in photosynthetic rate. Among other effects upon photosynthesis is a disturbance in starch metabolism leading to reduction in starch content.

Effects on translocation of nutrients and water.

For simplicity, transport will only be considered in terms of vascular tissues and across membranes. Nutrients and water are transported upwards in the xylem and downwards in the phloem, and the loading of these systems is facilitated by transport across membranes. The overall effect on translocation, of biotrophs and necrotrophs is similar, although the mechanisms are rather different.

indirect effects on translocation

Biotrophs tend to have indirect effects upon translocation and membrane transport, by altering stomatal aperture or by slowing root or shoot growth. Often these subtle effects tend to divert photosynthates to the region of infection rather than to the plant meristems.

Necrotrophs have more direct effects by killing plant material, often via toxin secretion. This, of course, alters the transport capacity of the affected tissue, and this in turn alters the translocation pathway.

tyloses

A particular group of pathogens called the vascular wilts spread through the plant in the vascular system, blocking the transpiration stream, and hence causing wilting. Ironically, the plant can exacerbate wilting through its defence response, which is to block the path of the fungus by producing tyloses in the vascular system. (The defence reactions of plants are described in greater detail in the BIOTOL text, 'Defence Mechanisms'.

Effects on plant growth

Obviously, the overall effects of these physiological factors will be to reduce growth in diseased plants. The fact that the photosynthetic ability of a plant is reduced by pathogen attack will obviously reduce biomass accumulation, but pathogen effects on growth can be more subtle through their effects on plant growth regulator (plant hormone) balance. In particular, pathogenic fungi can upset levels of auxins, gibberellins, cytokinins and ethylene in the plant. These substances are involved in the control of growth patterns in plants, and may contribute to some of the changes in growth and development of specialised structures often encountered in host-pathogen interactions.

effects on growth regulator balance

SAQ 4.1

Three different species of fungi are found to be associated with a particular variety of crop plants. The fruiting bodies of fungal species A are commonly found on detached leaves but not on leaves still attached to plants. The fruiting bodies of fungal species B are found on the leaves of dead plants or parts of plants that are still anchored in the ground. The fruiting bodies of fungal species C are found on distorted but living leaves of the host crop plants.

Identify as far as possible whether or not species A, B and C are biotrophs, necrotrophs or saprophytes, giving reasons for your selection.

4.3.5 Resistance to pathogen attack

Having talked about how pathogens enter a host plant, and the physiological effects manifested on colonisation, we have to ask what a plant can do to minimise damage from a pathogen. Such a process is generally called resistance.

You must note that most pathogens have a very limited host range, so that, except in rare cases, one pathogen species affects a single or, at most, only a small number of host species. Plant response can be seen as a graded response:

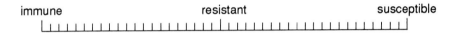

Two resistance strategies are utilised by plants against pathogens, both often acting in tandem:

- passive resistance - resistance due to the already existing state of the potential host plant;

- active resistance - resistance mechanisms activated by the plant directly in response to the presence of the pathogen.

It is important to realise that these resistance mechanisms can occur at any stage in the infection cycle.

Passive resistance

In general, passive resistance depends either on plant structure or on the presence of inhibitory chemicals.

Plant structure

Structural resistance tends to be exerted largely at the spore germination and penetration stages. Often leaves and other plant structures are shaped so that water droplets containing spores are unable to stay on the leaf. This can be seen as a form of *avoidance of infection*, since the plant might be susceptible to the pathogen if the spore-containing water droplets were able to remain on the leaf.

In some cases, cuticles may be physically tough enough to prevent or slow fungal penetration and colonisation, perhaps through the presence of silicon, or from lignification or suberisation.

Chemical resistance

saponins
catechols
phenolics

Many chemicals are inhibitory to pathogen growth. If these are pre-existing in plant tissue prior to pathogen attack and have the effect of slowing or stopping pathogen growth, then these are part of passive resistance. Examples of such inhibitory compounds include saponins, catechols and phenolics. Note that in some cases, although inhibitory compounds may be present in uninfected plant tissues, the concentration of these compounds may be altered by infections. In other words, although these compounds form part of the passive resistance, the stimulation of their synthesis in infected tissues indicates that, in some cases, they are also part of the active resistance.

Active resistance

Active resistance mechanisms are direct responses to pathogen attack, the plant attempting to slow or halt the spread of the pathogen. These mechanisms are described in detail in the BIOTOL text, 'Defence Mechanisms'. Here we will outline the main principles. We can list these as:

*   **production of physical barriers** - simply, the plant at the site of infection produces a barrier to limit or slow the growth of the fungus. Mechanisms include the formation of cork layers, the development of abscission layers so that the infected plant material can be shed, the deposition of non-degradable material in the path of the fungus. These include papillae and lignintubers in aerial tissue, and the formation of tyloses which serve to block vascular tissues in the paths of vascular wilt fungi. These structures generally contain cellulose, callose and lignin;

*   **production of chemicals** - chemicals can be secreted which are directly toxic to the pathogen. These are generally termed phytoalexins, and are thought to be produced by the host in response to the presence of the pathogen. A wide range of such chemicals have been isolated;

*   **hypersensitivity** - in certain host:pathogen interactions, particularly in cereals where the pathogenic organism is a biotroph requiring the presence of living cells, hypersensitivity is an important active resistance mechanism. This response is elicited on penetration by the pathogen and involves the death of plant cells surrounding the point of penetration. Without living cells, the pathogen is unable to mount a successful invasion. Because only few cells die, the infection does not severely impair plant performance.

4.3.6 Plant breeding for disease resistance

Although fungicides are a commonly used method to control plant disease, plant breeding offers an effective, economic and environmentally friendly method to control plant disease. It has long been appreciated that disease resistance is controlled at the genetic level. There are thought to be two forms of gene resistance:

*   major gene resistances;

*   polygenic resistance.

Major gene resistance

gene-for-gene
hypothesis

This is also called monogenic or race-specific resistance, where plants containing such a gene are resistant to a specific race of a pathogen. It often occurs as a hypersensitive response. Because only a single gene is involved, if the pathogen alters its virulence then the host's resistance will be overcome. This has lead to the suggestion that the host's resistance genes and the pathogen's virulence genes are closely related, the relationship being called the gene-for-gene hypothesis.

Polygenic resistance

Where more than three genes control resistance, this is termed polygenic resistance. Such genes tend to act in an additive manner and are not race specific. They are often termed partial resistance, and tend to have quite subtle physiological effects such as reducing infection frequencies, latent periods and reducing sporulation.

Major gene resistance is much easier to incorporate into crops in plant breeding programmes as it is simply inherited, but is more easily overcome by the pathogen.

gene-for-gene hypothesis

race-specific resistance

The suggestion that there is a linkage between a host's major resistance gene and a pathogen's virulence gene has been called the 'gene-for-gene' hypothesis. It is thought that if there is recognition between the gene products of the resistance and virulence genes, then the reaction between the host and pathogen is incompatible (ie resistance). If there is non-recognition, then infection can take place. This gives the idea of race-specific resistance.

The presence of major gene resistance in crop varieties does give problems, if reliance is placed on single major genes. As the pathogen evolves, it can overcome the resistance in the plant, giving a likelihood of a severe disease outbreak. Many plant breeding programmes try to combine a number of major resistance genes in order to make the host's resistance more durable.

4.3.7 Control of plant disease

There are several strategies that may be employed to control plant diseases. These include modifying cultivation practices, using resistant varieties or the use of chemical agents.

Cultural control

crop rotation

Cultural practices can influence the severity of the outcome of a disease outbreak. In particular, a knowledge of diseases' life cycles is important, giving the longevity of disease viability during resting stages, and the presence of alternative hosts. Rotations can form a powerful control method, giving a gap between planting of susceptible crops to reduce disease inocula sizes. Some diseases are able to remain in soils for many years, such as club root of brassicas, or wart disease of potatoes. Crops should not be planted until the resting inocula has died out. Weeds which could be alternative hosts and allow overwintering of spores should be controlled. Crop residues which could also act to carry-over disease to the next year should be destroyed or removed.

Resistant varieties

We have talked in some detail about plant resistance. The use of varieties resistant to problem diseases forms one of the most effective and economic control measures.

Chemical control

Chemicals, particularly fungicides, are available for the control of plant pathogens, with the exception of viruses (plants should be destroyed if possible). Fungicides can either protect against fungal attack, or destroy the disease organism in the plant. It is beyond the scope of this chapter to look at the myriad of fungicidal chemicals available, but when selecting a potential fungicidal chemical, the following points should be considered:

- toxicity: high level to fungus
                low level to plant;
- development of fungal resistance to chemical;
- environmental impact;
- ease/safety of application;
- cost.

| SAQ 4.2 | Re-examine the life cycle of the black rust disease-causing organism *Puccinia graminis* (see Figure 4.4) and suggest measures to control the disease. |

| SAQ 4.3 | Re-examine the life cycle of the late blight disease-causing organisms of celery (Figure 4.3) and suggest measures to control the disease. |

4.4 The effects of pests on crops

By definition, a pest is an organism which causes damage at an economically significant level to crops or animals being utilised by Man.

For our purposes, we will confine our discussion to pests of crop plants, although insects such as screw worms, warble flies and the tse-tse fly are devastating pests in animal production systems.

∏   Why do you think that we regard fungal pathogens as parasites of crop plants, whereas we consider insect pests to be predators of plants?

In general, damage caused by pests occurs due to their feeding habits, either as they consume plant material, or spread disease whilst feeding. Their effect may be solely to reduce biomass in the crop and hence reduce the final economic yield, or they may have more subtle effects on the end use of the crop by reducing the visual attractiveness of the product or by reducing palatability.

Pests are found amongst a great range of the phyla of the animal kingdom. A brief summary of these is given in Table 4.2.

	Group	Examples
invertebrates	nematoda	eelworms
	gastropoda	slugs and snails
	arthropoda	insects, beetles,weevils etc,
vertebrates	aves	birds
	mammalia	mammals

Table 4.2 Phyla containing the major pests in crop production.

Note that pests include both vertebrate and non-vertebrate examples. In practice, the most important group of crop pests are the Arthropods, and this group is described in further detail in Table 4.3.

Group		Examples
Insecta	Collembola	springtails
	Hemiptera	plant bugs, aphids, capsids
	Thysanoptera	thrips
	Lepidoptera	butterflies, moths
	Coleoptera	beetles
	Diptera	flies
	Hymenoptera	wasps, sawflies
Arachnida	Acari	mites
Crustacea	Isopoda	woodlice
Diplopoda		millipedes
Symphyla		symphilids

Table 4.3 Groups of arthropods which contain major plant pests.

The list given in Table 4.3 is quite extensive. We anticipate that you will be able to think of many specific examples.

As with plant diseases, the development of knowledge of the life cycles of plant pests can be of great value to farmers in enabling strategies to control the damage caused by such pests. The life cycles of this enormously diverse group of organisms are, of course, also quite diverse, so it is difficult to generalise. We will, however, give a couple of examples.

life-cycle of
Aphis fabae

Aphids are very common and cause damage to a wide range of crops. Black fly (*Aphis fabae*) are often found on beans and the closely related green fly are commonly found on roses. They cause damage not only by feeding on the plant sap, but also by being vectors of a variety of viruses. Here we will briefly describe the life cycle of *Aphis fabae*.

The life cycle of this species is illustrated in Figure 4.7. The attack on bean crops begins in early summer by the migration of winged females from their winter hosts (such as the spindle tree, *Euonymus*, and the guelder rose, *Viburnum*). On landing on the bean plant, these winged females produce young parthenogenetically at a rate of two or three per day. The first generation consists of wingless females which mature within about three weeks and produce winged females. These colonise further plants. Thus the winged females migrate and establish fresh colonies whilst the wingless females increase the size of the established colonies.

In autumn, winged females fly to the winter host where they produce a further generation of winged females. At the same time, winged males are produced on the summer host. These migrate to the winter host where they mate with the winged females. The females lay small black eggs in crevices and cracks in the winter host's bark. The eggs hatch in spring to produce winged females which migrate to the summer hosts.

This species is not confined to a single host. The summer generations can feed on a wide variety of crops (for example beans and beet) and a variety of weeds. This diversity of potential hosts makes the pest more difficult to control. It is usually achieved by using contact or systemic insecticides.

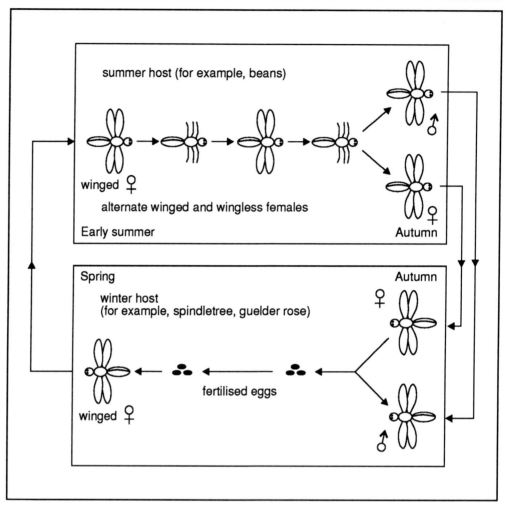

Figure 4.7 Life cycle of *Aphis fabae*.

life-cycle of
Pieris brassicae

The cabbage-white butterfly (*Pieris brassicae*) is another well known pest. The adult butterflies appear in May and, after maturing, the females lay eggs in batches (10-100) on the undersides of the leaves of various brassicas. The eggs hatch within about 10 days and the emergent caterpillars feed voraciously on the leaves. After several (3-4) months, the caterpillars pupate. This is usually accomplished by the caterpillar migrating to some raised area protected by an overhanging ledge. There they spin a small quantity of silk by which the caterpillars are suspended. Then they undergo their final moult to form a hard-coated pupa (or chrysalis).

Within the pupa, metamorphosis takes place over a period of 2-3 weeks. Then the imagines (imagoes) emerge. This generation of adults produces eggs in late summer/early autumn. The caterpillars mature in autumn and they overwinter as pupa. The imagines emerge in the following May (see Figure 4.8).

The caterpillars (larvae) do considerable damage to brassica crops.

∏   See if you can think of ways of reducing this damage.

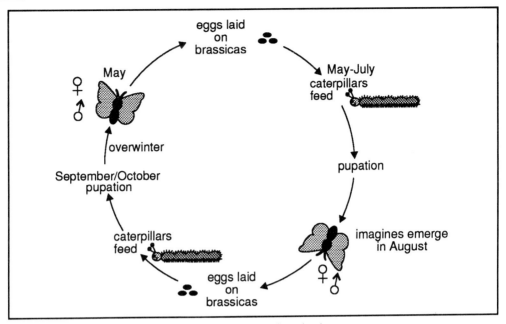

Figure 4.8 Life cycle of the cabbage-white butterfly (*Pieris brassicae*).

cabbage-white
butterflies
difficult to
control

Cabbage-white butterflies are difficult to control except by using chemical means. Conventional methods used on a garden scale include squashing eggs and hand-picking larvae. Such methods cannot be economically applied to the field. Instead chemical insecticides are used. Because of the environmental consequences of using recalcitrant chemical sprays, many attempts of using biological agents have been made. These caterpillars are not much favoured by birds. They are, however, extensively parasitised by ichneumon flies which lay their eggs in the caterpillars.

The ichneumon flies, however, only slowly kill their hosts and the cabbage-white caterpillars continue to feed long after initial parasitisation. Thus this approach to control is not very satisfactory. Alternative strategies include using natural diseases (bacterial or viral) of the cabbage-white butterfly. The application of biotechnology to produce more environmentally-friendly crop protectants is described in the BIOTOL texts, 'Biotechnological Innovations in Crop Production,' and, 'Biotechnological Innovations in Environmental and Energy Management', so we will not go into the technical details here.

winter moth

Sometimes knowledge of the pest's life cycle indicates an obvious point at which the pest might be controlled. For example, the winter moth is a major pest of apple orchards. The females are wingless. Emerging from their subterranean pupa, they crawl up the trees to mate with the winged males. They lay their eggs in the finer branches and twigs. The caterpillars feed on the leaves and, when mature, spin silk threads to lower themselves to the ground. They bury themselves and pupate over winter. The caterpillars (larvae) of this moth may completely strip the leaves from the branches in April-May.

Π   Suggest a simple way of controlling the winter moth without using toxic chemicals.

In practice a common method of control is to place sticky bands of wax around the tree trunk. The wingless nature of the females means that they must climb the trunk in order to lay their eggs in the appropriate places. The 'sticky bands' entrap the migrating females and the life cycle of this species is, therefore, disrupted.

We will discuss control measures again in Section 4.4.3.

4.4.1 Economic injury

As in the case of plant disease and plant competition, whether an organism is regarded as a pest is determined by economic factors.

Crucially linked to economic damage is the population level (density) of a species within a susceptible crop. Obviously, if a population exceeds a given level, then the farmer may be the subject of potential income losses from yield reductions due to that pest, unless control measures are taken.

This consequence has led crop protection experts to analyse pest populations in more detail, and they have identified 3 particular population densities as keys in the assessment of pest effects. A knowledge of these levels by the farmer is essential for the economic application of control measures. These three levels are:

- the population equilibrium;

- the economic injury level;

- the economic threshold level.

Population equilibrium is the mean long-term population density of a pest, unaffected by any control measures. The population will fluctuate around this value dependent on any external biotic and abiotic factors affecting population levels.

Economic injury level is the pest population density at which damage occurs at a level where control measures become economically favourable.

Economic threshold level is the lowest pest population density where it becomes economic to institute control measures. This is generally marginally less than the economic injury level, in order to prevent the pest density from reaching the economic injury level.

From these concepts, you can see that the objective of any pest management programme is to keep the pest population below the economic injury level. These concepts are probably best illustrated by some examples which are given in Figure 4.9 a-d.

Π Examine the examples in Figure 4.9 and try to assess the seriousness of each of the pest examples.

non-pests, occasional pests, perennial pests and severe pests

Obviously, yellow woolly bear on maize never reaches population levels high enough to cause injury at an economic level, so is a non-pest of maize. Green cloverworm on soyabean does sometimes reach population levels that cause economic damage to soya, but not regularly. Such pests are termed occasional pests. Colorado beetle is a perennial pest of potato, continually requiring regular control treatment to keep populations below the economic injury level. In some cases, such as the apple codling moth, a pest is severe, because the population equilibrium is above the economic injury level. Only rigorous control measures can reduce population levels and maintain a modified population equilibrium below the economic injury level (see Figure 4.9d).

You must remember that these criteria are economic. The effect of a pest may only be visible in terms of a farmer's balance sheet. For example, a pest may have no other effect but to delay harvesting of a crop, without reducing yield or quality. Later prices at market tend to be lower than the prices for early crops; for a late harvested crop a farmer may receive less than for the same amount of an early harvested crop.

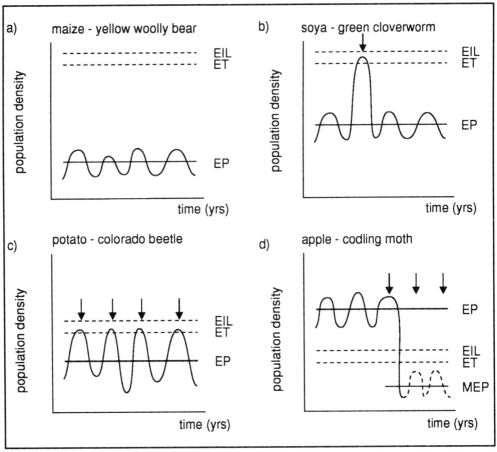

Figure 4.9 Examples of population densities of various potential pests on a variety of crops. EP = population equilibrium; ET = economic threshold; EIL = economic injury level; ↓ = pest control intervention; MEP = modified population equilibrium. (Adapted from Mathews Pest Management, Longman, 1984).

Although we have talked about a population equilibrium, such equilibria are themselves dynamic and can change. Figure 4.10 gives 2 theoretical examples to represent the extremes of population change.

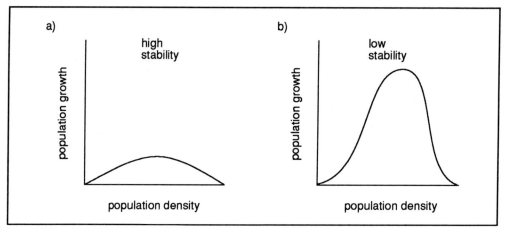

Figure 4.10 Theoretical examples representing the extremes of population change; a) pests with stable habitats, b) pests with unstable habitats (see text for discussion).

The population with low habitat stability shows enormous changes in population growth. Pests in this situation tend to be opportunists, small, mobile, and migratory, with populations going through boom and bust cycles. The classic example of this is the periodic swarming of locusts. At the other extreme are pests with highly stable habitats where populations stay fairly stable, although often filled to their carrying capacity. Pests of these habitats tend to be of larger size, less migratory and tend to have longer generation times. You must note that the disturbed nature of agricultural habitats tends to make them fairly unstable habitats.

We have developed the idea that the level of economic injury is dependent upon the level of infestation ie the population density. This idea can be extended by looking at a generalised scheme to relate yield to population density as shown in Figure 4.11.

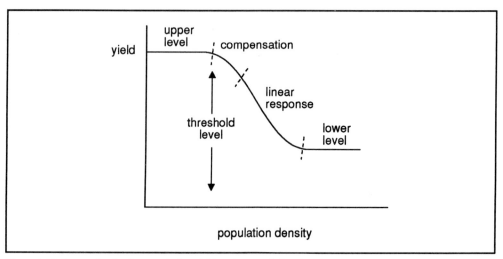

Figure 4.11 Generalised scheme relating yield to the population density of a pest.

This is very much a generalised response and does not represent every plant-pest interaction. A few other examples of responses are shown in Figure 4.12.

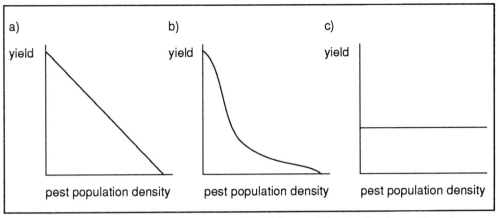

Figure 4.12 Three types of relationships between yield and the population density of a pest (see also Figure 4.11).

☐☐   Which of the relationships shown in Figure 4.12 indicate that the loss in yield is independent of the population density of the pest?

You should have identified relationship c).

4.4.2 How do pests damage plants?

☐☐   Make a list of how pests may cause reductions in crop yield and then check your list with the discussion given below.

Direct damage

Direct damage involves the loss of plant parts which would have directly contributed to the economic yield of the crop. Examples would include the complete loss of crops by such pests as locusts or army worms, boring insects in fruit crops, or the consumption of grassland by rabbits. Most leaf feeders cause direct damage by consuming the harvested component of the crop.

Indirect damage

Where the action of the pest does not directly consume yield, but causes a subsequent reduction in yield, this is termed indirect damage. Most leaf consumers effectively reduce the total leaf area of a plant. If the leaves are not the final product, then a reduction in the photosynthetic area contributing to yield will occur and the yield will be lower. For example, the consumption of the flag leaf of wheat will reduce the ability of the plant to fill grain with photosynthetically-derived assimilate. Indirect damage also occurs in the root environment, leading to a reduction in the plant's ability to take up water and nutrients. Examples of this type of damage would include the action of nematodes, cutworms or leather jackets on roots.

Pests as transmission vectors

Pests such as insects tend to be highly mobile, and can act as carriers of pathogens. They, therefore, have indirect effects as transmission agents of these pathogens. For example, aphids are very important vectors of a huge range of virus diseases through their ability to feed directly on plant sap. Examples of fungal diseases carried by insects include Dutch elm disease (*Scolytus* beetles) and coffee rust (coffee berry borer).

Table 4.4 gives a range of damage caused by pests on plants:

	Type	Damage	Plant effect
direct	biting mouthparts	consume leaves	reduction in total photosynthetic capacity
		tunnel into stem (borers)	weaken stem and reduce translocation
		ring the bark	restrict translocation
		consume growing points	growth distortion
		cause fruit fall	
		attack flowers	reduce seed yield
		attack seed	reduce seed yield
		attack roots	restrict uptake of water and nutrients
	piercing/sucking mouthparts	consumption of sap	reduce plant vigour, causing wilting
		inject toxin	growth distortion , necrosis
		provide entry for pathogen	incidence of disease increased
indirect		presence of insect	make cultivation more difficult
			delay harvest
			reduce palatability
			reduce nutritional quality
			transmission of disease

Table 4.4 Types of damage caused by pests on plants.

4.4.3 Methods of pest control

∏ List 3 or 4 approaches that could be used for pest control.

The approaches used to control pests are similar to those used to control disease, and can be divided into 4 main areas: pest exclusion; pest avoidance; use of chemical protectants, eradication.

Pest exclusion: methods designed to keep the pest out of a particular area or crop (for example quarantine).

Pest avoidance: use of sites free from infection. Use cultural methods to prevent infection, and use resistant varieties.

Chemical protectants: use of chemicals which confer protection from attack.

Eradication: use of chemicals to control the outbreak of an infection.

Also under eradication we can cite the use of biological agents (for example viruses) that may be used to eradicate pests. The biological control of pests is becoming increasingly important. This is partially a reflection of improvements in the construction of biological control agents using contemporary molecular biological procedures and partially a reflection of the increasing concern being expressed over the environmental consequences of using non-biodegradable biocides.

From these broad principles, a range of appropriate control measures have been developed. We will provide a summary of these.

Legislation/sanitation:

- compulsory quarantine;

- prevention of spread through use of appropriate sanitation measures.

Cultural control:

- ensuring maximum plant growth and photosynthetic efficiency;

- healthy plants resist pest attack better;

- adjustment of sowing and harvesting times to avoid the presence of the pest;

- adjustment of cultivation practice (for example deep planting);

- removal of secondary hosts;

- use of resistant cultivars.

Biological control (the use of a living organism as a control agent):

- introduce (or boost levels of) parasites, diseases, predators of the plant pest;

- male sterile programmes;

- disruption of behaviour of the pest by, for example, application of pheromones.

Chemical control:

- use of repellents/anti feedants, ingestive poisons, contact poisons, systemic poisons.

Although we have largely examined pests of growing crops, the principles we have considered also apply to stored crops. From the following data, determine under what conditions the grain mite *Acarus siro* is a pest of grain stores, and suggest some potential control measures.

Relative humidity (%)	Temperature (°C)							
	0	5	10	15	20	25	30	35
95	x	✓	✓	✓	✓	✓	✓	x
90	x	✓	✓	✓	✓	✓	✓	x
70	x	✓	✓	✓	✓	✓	✓	x
65	x	✓	✓	✓	✓	x	x	x
62	x	x	✓	✓	✓	x	x	x
60	x	x	x	x	x	x	x	x

x = grain mite not present
✓ = grain mite present

4.5 Effects of plant competition on crops

Crop yields are not only affected by the ravages of pests and diseases, but also by the effects of neighbouring plants. You must remember that crop productivity reflects yields not from single plants but the combined production of groups of plants, namely the crop canopies.

Competition can be defined as the interaction between neighbouring individuals in a plant or crop community in order to obtain mutually required resources from the environment that may be limiting.

We should also define what is meant by a plant population. This varies with crop species and largely depends on the morphological structure of the crop plant. Crops such as maize and sugar beet are relatively simple structures based on a single stem, and, as such, the individual plant represents a unit of population. Where more than a single stem is produced from a single seed (or, perhaps, a potato tuber), the situation is less clear cut. For example, cereal plants tend to produce tillers only some of which will produce useful yield. In this case, the number of ears per unit area, rather than plants per unit area may be a more meaningful measure of plant population.

∏ How do you think that plants within a crop canopy are affected by neighbouring plants?

Competition by plants with their neighbours reflects a struggle for the resources available within the production environment. You should now be familiar with the physical resources in the environment that contribute directly to plant and crop growth, they broadly fall into three areas:

- water;
- a range of soil nutrients;
- light and CO_2.

Where the supply of one of these parameters is reduced to a sub-optimal level, plant growth and hence crop yield may be reduced. Where such a factor is sub-optimal due to the presence of a neighbour, this is termed competition.

Π There are generally considered to be 2 forms of plant competition, depending upon the species found in the crop canopy under consideration. Write down how these two forms might differ.

intraspecific competition

Since a plant's neighbour may cause competition, the nature of the neighbour affects the type and extent of competition. Neighbouring plants may be of the same species, as would be expected in a typical monoculture crop, and would have similar physiological requirements to the crop environment. Such competition is generally called intraspecific competition. Within such monocultures, considerable effort has been made by agronomists to optimise the density and spacing arrangement of plant populations within crops, so that efficient use of resources is made but without yield being excessively restricted by plant competition.

weeds

interspecific competition

A second form of competition arises from the effects of non-crop species within the stand of the desired crop. Such species are generally termed weeds, and in modern agricultural systems are considered undesirable as they remove resources which could otherwise be utilised by the growing crop. Such competition has been termed interspecific competition.

With increased pressure to reduce pesticide and herbicide inputs into agricultural systems, an understanding of how weeds interact with their crop plant neighbours is of increasing importance.

4.5.1 Intraspecific competition

Within a crop stand, plants compete for resources from the aerial and soil environments. Although the CO_2 concentration is markedly sub optimal especially in high density situations, high density crops are more likely to be limited by light than by CO_2 concentration. As the canopy develops, plants grow taller and leaf density increases. At some stage in development, lower leaves may become shaded by upper leaves and the rate of photosynthesis is reduced. Furthermore, leaves from neighbouring plants may overlap with similar results. This leads to an inefficient canopy system in terms of carbon acquisition, which may reduce yield.

Within the root environment, densely packed plants within the canopy may restrict the volume of soil available to be occupied by a single root system. This will obviously reduce the amount of nutrients available to each plant, and will also reduce water availability. Densely packed plants tend to be more prone to drought and generally require higher applications of fertiliser to maintain yield.

∏ Consider how plant competition might vary as a crop grows. Do you think that plant competition will be the same or different in a stand of barley at the following growth stages:

- seedling;

- anthesis (flowering);

- approaching harvest.

Obviously as individuals develop, competition for light and nutrients will tend to increase. In stands of seedling plants, with neither root or leaf systems of individuals meeting, competition will be low. As individuals develop (that is, get larger) competition will inevitably increase. So, in our example, barley will suffer more competition when approaching harvest than during anthesis. At higher densities, where plants are initially closer together, leaf and root systems will meet earlier in the season, and hence competition will start earlier.

Optimising plant competition within crop stands

Agronomically, for maximising yield, it is necessary to choose a plant density which avoids the inefficient use of resources at low populations, but avoids the effects of excessive competition at high populations. Considerable effort has been made by agricultural institutes and advisory services to determine sowing rates of crops to give the most efficient crop canopy.

In general, 2 types of crop response to plant density occur. These are asymptotic relationships and parabolic relationships. An asymptotic relationship is shown in Figure 4.13.

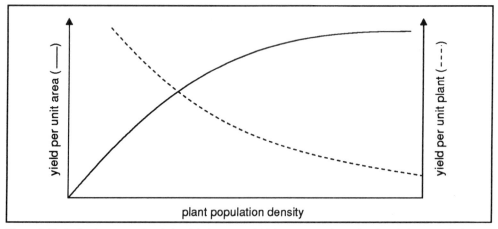

Figure 4.13 Stylised asymptotic relationship between yield (per area and per plant) and plant population density. (———) = yield per area; (------) = yield per plant.

A large number of temperate crops show an asymptotic relationship between plant density and yield. This is particularly true of forage crops, but is also true for some grain crops. You will notice how the total yield per unit area becomes more-or-less constant at high plant population densities. Thus, it is pointless to increase the plant density above this point as this would be both wasteful of seed and would incur the cost of the wasted seed without generating any increase in saleable yield. If plant densities were established below this level then the maximum yield per unit area would not be achieved.

Figure 4.14 shows a parabolic relationship between yield and plant population density.

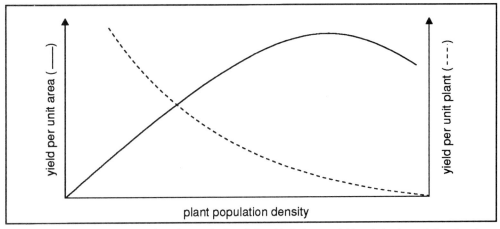

Figure 4.14 Stylised representation of a parabolic relationship between yield and plant population density. (———) = yield per area; (------) = yield per plant.

Crops showing parabolic relationships between yield and plant density, show a decline in total yields above a particular plant density. Note that in comparison to the asymptotic relationship, individual yield per plant declines quite markedly at high plant densities. An interesting example showing the complexity of the situation is the effect of seed rate on potato yield. The relationship with total yield is asymptotic but, at high plant densities, many of the potato tubers are too small for commercial use and the utilizable yield shows a parabolic relationship.

The arrangement of individual plants within crop stands

So far we have considered crop stands in terms of plant density (the number of plants per unit area). Obviously, within this area plants can be arranged in different fashions, dependent upon the variables:

- row width;

- distance apart within row.

In modern agricultural systems, sowing and planting are performed mechanically so that the above spacings determine plant density. In more primitive agricultural systems, seed is often broadcast, giving much less regularity in the arrangement of plants within a stand. The balance between row width and distance apart determines the shape of the area occupied by a single plant.

Consider the two following arrangements of plants.

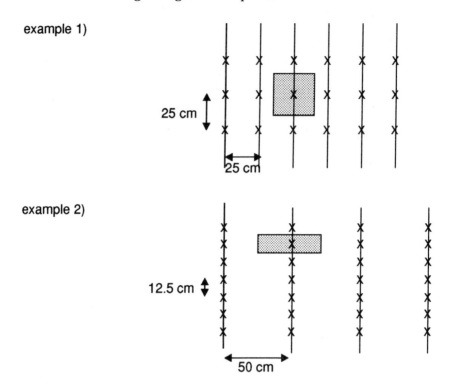

example 1)

25 cm

25 cm

example 2)

12.5 cm

50 cm

In the above examples, plants occupy the same area (625 cm^2), although the 'footprints' of the plants differ. In example 1 with a square footprint, we might expect competition to be marginally lower and yields higher. Nevertheless, wide row widths make cultivation easier, ie for the application of fertilisers and sprays during growth, and hence are generally preferred. It has been found that the plant arrangement has, however, little effect on crop yields at plant densities typically used in most crops.

SAQ 4.5

From the following data for some yield components of spring barley at different sowing densities, calculate the final grain yield and plot yield against plant density. Comment on the effect of plant density on yield and yield components in this case.

seed rate (kg ha^{-1})	86	118	146	183	218
plant density (x 10^6 ha^{-1})	1.99	2.74	3.29	4.35	5.21
ears/plant	2.96	2.26	1.99	1.60	1.34
ears/ha (x 10^6 ha^{-1})	5.89	6.19	6.55	6.96	6.98
grains/ear	23.8	23.2	22.8	22.0	21.7
weight of 1000 grains (g)	36.9	35.7	36.6	35.7	35.2
Yield = ?? (t ha^{-1})					

4.5.2 Interspecific competition

Not only competition between plants of the same species may occur within a crop stand, unrelated species may also be competitors. In some agricultural systems crops are sown not as monocultures as in Western agriculture, but as mixed cultures. The objectives of such a system is to more fully maximise the environment's resources over the course of the growing season. Nevertheless, interspecific competition is normally considered as the effect of a non-economically useful species on crop yield; namely the effect of weeds.

∏ What effects do you think that weed species may have on the growth and yield of crops?

The competitive nature of weeds is essentially similar to that shown by neighbouring plants undergoing intraspecific competition and arises from their requirements for water, nutrients and light. The effect of the presence of weeds is generally to reduce yield. The factors affecting interspecific competition are given in Table 4.5.

crop species	growth habit	growth rate/vigour	earliness
weed species	growth habit	growth rate/vigour	earliness
weed density	density	weed duration	
soil status	nutrients	water	

Table 4.5 Factors affecting inter-specific competition levels and crop yields in crop monocultures. (See text for discussion).

∏ Is an early variety of crop plant likely to be at a competitive advantage or disadvantage to a late variety of weed species?

You should have concluded that it would have an advantage as it would grow first and fill the leaf canopy thus reducing the amount of radiation received by the weed plants. Thus the weed plant would be, most likely, light limited. In a similar way, an early variety of weed would have an advantage over a late variety of crop plant. However, matters are not usually so straightforward.

growth habit of weed important

The relative growth habits of crop and weed are important. A prostrate weed, such as knotgrass, will not be a serious competitor for light in a tall crop such as wheat, whereas taller weeds are particularly serious competitors in short, slow-growing crops such as sugar beet. Sugar beet is sown in March and April and, therefore, it is particularly vulnerable to weeds which exhibit rapid early growth leading to light competition. Fat hen (*Chenopodium album*) is a typical example of a weed which may affect sugar beet yields.

morphology of root systems of weeds are important

Additionally, competition for water and nutrients is important. Most weed species are able to form deep tap root systems which are better at exploiting the available soil volume than the fibrous root systems of crop species. It is thought that prostrate weeds are particularly important competitors for nutrients, especially nitrogen, whereas tall-growing species are the main competitors for light.

Perhaps the most important factors are weed density and duration. The extent of competition is greater at higher densities, and where weeds are dense, competition tends to occur earlier.

∏   A particular variety of plants is described as being faster germinating than related strains. Is this property likely to increase or decrease this variety's competitive ability.

germination rate and its effect on competition

It is likely to increase its competitive ability. Fast germination means that this variety is likely to emerge early and to establish the canopy before the weeds emerge. By germinating early in spring, these varieties are able to form the leaf canopy before the weeds can become established. This is, however, a great simplification of what may happen in practice. The competitive ability of a crop depends upon a wide variety of factors such as growth rate, leaf morphology, avidity with which it can absorb nutrients, and root morphology.

In Figure 4.15 we have provided some data relating to the effects of barnyard grass (*Echinochloa anus-galli*) on maize. Note that the higher the density of the weed, the greater the effect on yield.

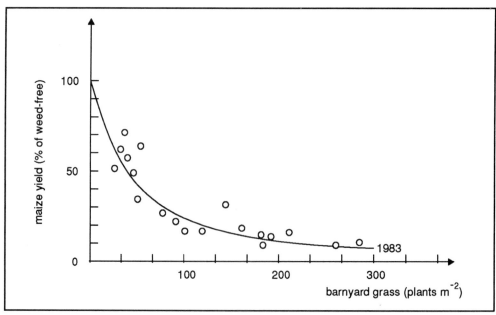

Figure 4.15 Shoot biomass of maize relative to that of weed-free maize as a function of weed density for two seasons. (Based on the data of Kropff, KJ, Vossen, *et al.* 'Competition between a maize crop and a natural population of *Echinochloa anus-galli*'. Netherlands Journal of Agriculture Science 32, pp 324-327 (1984).

The time delay between crop and weed emergence is also important. In Figure 4.16, we have provided some data relating sugar beet yield to weed density as a function of the delay between crop and weed emergence. The weed referred to in these data is fat hen (*Chenopodium album*).

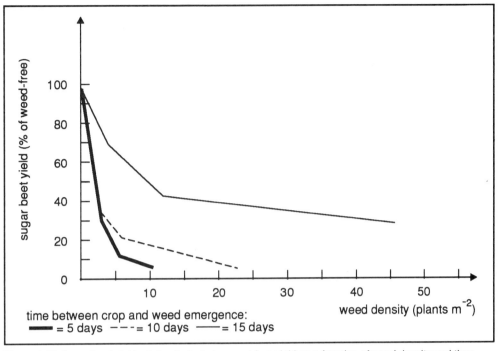

Figure 4.16 Sugar beet yield relative to that of a weed-free yield as a function of weed density and time between crop and weed emergence. Data stylised from that of Bos, B and Wallinga, J 'Quantificering en validatie van een eenvoudige onkruid-schade relatie; onderzoek concurrentie tussen het gewas suikerbiet en 3 onkruidsoorten met veld-en simulatie experimenten. *Onderzoekverslag, landbouwuniversiteit Wageningen* (1991) (see text for discussion).

As might be anticipated, the sugar beet yield decreased with increasing density of fat hen. However, the high weed density may be counterbalanced by a low individual plant weight of the weed if there is a long delay in the emergence of the weed. The resulting differences in their respective starting weights favour the less densely planted but larger sugar beet plants. This may lead to a high competitive ability in exploiting growth factors like light, water and nutrients and hence a less pronounced yield reduction.

competitive ability

This type of data shows how difficult it is to predict the exact consequences of weed infestation on crop yields The two key points to remember are that, in general:

- higher seed densities will cause the greatest reduction in yield;

- the earlier the weed emerges (relative to the emergence of the crop), the greater will be its probable effect on crop yield.

timing of weeding operations is critical

It should be self-evident that the timing of weeding operations (for example mechanical hoeing, spraying, etc) is also critical. If it is done too early, freshly germinated weeds may be able to establish themselves at the expense of the desired crop. If done too late, mechanical damage to the canopy may occur and this may allow diverse organisms to gain access into the crop plants. Agriculturalists are beginning to develop crop simulation models that can play an important role in effective weed control.

The presence of weeds in crops does have a range of other consequences, apart from the direct effects on competition. Competition itself can reduce the quality of the crop product, either by reducing the size of the yield unit (for example, grain size), or due to the presence of weed seed which can make crops less suitable for processing (for example, for grain milling into flour). Sometimes, weed seeds may be poisonous, making human consumption of the yield product dangerous (for example, presence of corn cockle seed in grain). Additionally, weeds can hinder the use of machinery during cultivation. For example, mats of perennial weeds can slow ploughing or drilling through their effects of blocking tines or drills. During harvest, the presence of high weed densities can slow machinery and reduce yields.

Weeds also can act as alternative hosts for pathogens and pests. In particular they can allow the overwintering of pests and diseases, which are then ready to re-inoculate the following year's crop. Examples include couch grass (*Agropyron repens*) which can overwinter the 'take-all' disease organism, barberry which is an alternative host for black rust (*Puccinia graminis*) of wheat and a range of weed species which act as alternative hosts for nematodes and aphids.

4.5.3 Methods of weed control:cultural control

Although it is beyond our scope to look at weed control, especially chemical control, in detail, we should at least consider what approaches are available to the farmer to restrict competition from weeds. In general, weed control methods are divided into chemical control (ie the application of herbicides to kill or restrict weeds) or cultural control measures. Cultural control measures include:

- cultivation;

- crop rotation;

- roguing.

We will briefly examine each in turn.

Crop cultivation

mechanical and zero tillage systems

Cultivation involves the mechanical tillage of the upper soil horizons, and has been a feature of traditional agricultural systems for thousands of years. Recently, zero tillage systems have become popular in modern arable farming, where seed is sown directly into the stubble of the previous year's crop, without any tillage of the upper soil surface. Cultivation methods, such as ploughing, have the effect of burying seeds and any perennial structures such as rhizomes. This will tend to prevent reappearance of these weeds.

Repeated cultivation can stimulate weed growth as seeds are returned to the soil surface, although in the case of many perennial species, repeated cultivation can kill weeds through depletion of their carbohydrate reserves. With some rhizomatous weeds, such as couch grass, the continued chopping of the rhizomes can act to spread the weed. Nevertheless, in general, cultivation can successfully restrict weed populations, especially when combined with other control methods. Examples of weed densities after a range of cultivation methods have been used are shown in Table 4.6. Note that the figures vary enormously between plots and generally, ploughing reduces weed densities more than cultivation and cultivation is better than zero tillage at reducing weed density.

Cultivation Method	Density (plants m^{-2})
ploughing	13.3
repeated cultivation	98.6
zero tillage	142.3

Table 4.6 Mean blackgrass density at various sites, where different cultivation methods have been used. (After Harper, Principles of Arable Crop Production, Granada, 1983).

Rotation of crops

Although the use of crop rotation has declined in recent years as continuous cereal monocultures have become the dominant feature of arable farming, rotation allows an annual change of land use giving different times of cultivation and different sowing dates. This stimulates changes in the weed flora, which discourages any one species becoming dominant. The increase in the amount of herbicides that have been used has meant that the importance of rotation as a weed control strategy has declined.

Roguing

Where weeds are present at low densities in high-value crops, it may be economic to remove weeds by hand. This process is known as roguing.

4.5.4 Chemical control of weeds.

Chemical control methods involve the application of chemicals (herbicides) to crops. There are now over 100 chemical substances available in more than 500 formulations as weed control agents. Obviously, with the diversity of chemicals available, the choice of a particular herbicide for a particular application is complex. Here we will examine some of the underpinning principles which are applied in selecting suitable herbicides.

many herbicide formulations available

Chemicals have been screened, developed and formulated to give a whole arsenal of treatments to kill a wide range of weeds. Bearing in mind that there are hundreds of different weeds affecting a whole range of crops, chemicals have been developed for specific tasks. Herbicides can be selected according to a range of qualities and requirements including their specificity, rate of application, rate of action, time of application and their persistence.

Range of action: selective or non-selective

Non-selective herbicides kill all (or a large number) of plant species, and generally have to be applied prior to the emergence of the crop. Where total weed control is required (ie on industrial areas, roadways, or railways), these herbicides are designed to be persistent. Selective herbicides have a narrower range of action, and can be applied to weeds in non-susceptible crop species, post emergence.

Site of application: soil or foliar

This is usually either via the soil or via the leaves and depends on whether the herbicide is a contact herbicide or if it can be translocated within the plant.

Site of action: contact or translocated

Contact herbicides only act at the point of application and, therefore, tend to be applied to the foliage. Systemic herbicides can be transported within the plant and, therefore, can be applied at one point and be effective at another (ie shoot to root translocation).

Time of application in the crop's life cycle

We can divide herbicides into three broad categories according to the stage in the crop's life cycle in which they are used. These are pre-sowing, pre-emergence, and post emergence herbicides.

Persistence

The length of time that the herbicide stays active in the environment to which it has been applied is highly variable. Some are short-lived, some persist for a very long time (several years). Obviously persistent types are useful for clearing paths and roadways. Their use on arable land may present problems especially if a system of crop rotation is used. The crops used must all be insensitive to the persistent herbicide otherwise the herbicide resident in the soil will reduce/prevent their growth.

Recalcitrant (persistent) herbicides are of major environmental concern. They tend to get washed out into rivers and accumulate in organisms at the higher ends of the food web. Thus there are many efforts being made to reduce reliance on these types of agents.

4.6 Integrated pest management

From the grower's point of view, improving yields is a priority. Diseases, pests and weeds have drastic effects on potential yields (see Figure 4.1) and control of their agents has become one of the major activities in agriculture.

∏ Now that you have read Sections 4.1-4.5, how would you go about maximising yield? Do you think that there are any environmental consequences from your proposed strategy?

In the 1940s and 1950s, a range of chemical pesticides was developed (fungicides, insecticides, herbicides, etc) to give the grower effective control methods. At the same time, the world population grew rapidly, increasing the need to maximise food production. These chemicals formed very effective methods to close the gap between actual and potential yield, and great reliance has been placed on them.

Nevertheless, over the last 20 years, it has been recognised that these chemicals have a range of deleterious effects on the general environment, leading to their progressive banning, or restriction on their use. Additionally, the target organisms for many of these chemicals have become resistant to their effect. The development of a new pesticide now costs in excess of £25 x 10^6. In view of these factors, alternative methods of pest control are now being examined.

IPM

We hope that through your reading of this chapter you have come to the conclusion that there are a range of control options open to the grower. Making the best use of these options is known as Integrated Pest Management (IPM).

Integrated pest management is an overall strategy that makes use of an appropriate blend of biological, chemical, and cultural control methods.

This definition could be extended to include making use of appropriate forecasting methods.

The philosophy of IPM is generally held to try to minimise the environmental impact of the control methods employed. This is particularly so in the use of chemicals, where the approach steers away from the application of broad spectrum, prophylactic chemical treatments, to highly targeted, specific chemical control measures. In the future, it is highly likely that the use of chemicals will be increasingly restricted.

To set up a successful integrated pest management programme, some basic rules have been developed based on the experience gained as programmes have evolved. These are:

- an understanding of the biology of the crop is necessary. In particular a knowledge of the stages at which the crop is most vulnerable, and hence needs most protection;

- a knowledge of the key pests which cause economically significant damage to the crop is needed. These particular pests can then be targeted;

- an understanding of the biological and physical factors which determine populations of key pests is also required;

- there is a requirement for adequate surveillance and forecasting of pest populations;

- susceptible stages in the pest's life cycle, which are appropriate in the timing of control measures, need to be identified;

- flexibility; IPM relies on being able to institute control measures at short notice, rather than to a set, rigid pattern;

- minimisation of environmental impact. This prevents disruption of natural populations of predators and pathogens;

- maintenance of ecological diversity. A diverse ecosystem prevents the build up of dominant pest populations.

Obviously an integrated pest management programme is more difficult to manage than traditional pest control measures, because of the need to continuously monitor the agricultural ecosystem, and the need for rapid decision making. Nevertheless, IPM is likely to represent the future in pest management strategies.

Summary and objectives

In this chapter, we described the biotic factors which may reduce yields of crops. We focused on plant diseases, pests and competition between plants. The attack of plants by pests and diseases and the effects of competition are diverse and complex involving many different relationships. Here, we mainly dealt with the general principles and examined a few specific examples as illustrations of these principles.

Now that you have completed this chapter you should be able to:

- list the major groups of organisms which cause disease in plants;

- list the major groups of plant pests;

- describe the physiological processes of plants which may be affected by disease-causing organisms and describe the range of symptoms they produce;

- explain, by using suitable examples, how a knowledge of the life cycles of disease-causing organisms and pests can be used to control yield losses;

- distinguish between biotrophs, necrotrophs and saprophytes, disease-causing organisms and pests;

- distinguish between passive and active resistance and describe the main differences between monogenic and polygenic resistance;

- explain the major strategies that may be used to control diseases and pests and interpret data relating to various aspects of pest control;

- explain the differences between non-pests, occasional pests, perennial pests and severe pests and list ways in which pests may damage plants;

- distinguish between intra- and interspecies competition and explain how a knowledge of the effect of plant population density may be used to maximise crop yields;

- describe the strategies that may be used to reduce losses which may arise from competition from weeds and explain why it is difficult to predict losses arising from competition from weeds;

- explain what is meant by integrated pest management and list some of the basic rules which apply to it.

Greenhouse and growth rooms

Greenhouse and growth rooms

5.1 Introduction

control of
environment

The main aim of using greenhouses and growth rooms is to provide a protected environment which is controlled to varying degrees according to the proposed use. The reasons for providing this protection are numerous and varied, encompassing the need to provide fresh vegetables out of season, grow sensitive plants in otherwise intolerable climates, provide precise growing conditions for experimental purposes and the purely commercial enterprise of supplying large quantities of house plants and cut flowers all year round.

The degree and extent of control provided essentially determines the designation of the controlled environment as a greenhouse or a growth room chamber. A greenhouse is only another form of an environmentally controlled chamber characterised by the use predominantly of natural light and by the limits within which the conditions are constrained. The term 'greenhouse' covers a wide spectrum of structures from the remarkably simple overgrown cloche (basically a hemispherical plastic tunnel with rudimentary ventilation systems and shading/cooling with straw mats) to the all glass multispan, centrally heated, computer controlled glasshouses that are increasingly common in Northern Europe.

∏   What are the factors that can be controlled in these protected environments?

Table 5.1 illustrates the relevant factors and the degree of control that can be exercised over them compared to the natural environment.

	Field	Greenhouse	Growth room
temperature	no	some	yes
humidity	no	some	yes
irrigation	some	yes	yes
wind	no	yes	yes
nutrients	some	yes	yes
atmosphere	no	some	yes
day length	no	some	yes
repeatable	no	some	yes
uniform	some	some	yes

Table 5.1 A comparison of the extent of control over environmental factors possible in protected conditions. Adapted from Sanyo-Gallenkamp manual, 1993.

The style and degree of sophistication of the facility will depend upon a large number of factors, but will always, in the final analysis, come down to the proposed use and economic viability of the project. There is clearly little point in developing elegant greenhouse facilities, when the plants in question will grow perfectly well in an open field. There are a number of different types of greenhouse in use but to illustrate the basic requirement we shall concentrate on the standard form of greenhouse common in Europe and describe the factors that need to be considered when erecting and operating such a structure.

We shall see later how the conditions above are achieved and consider the consequences for plant growth. It is necessary in the first instance to describe the environmental factors involved and the way they are measured in the most appropriate fashion.

This is a long chapter, so do not attempt to study it all in one sitting.

5.2 Temperature

Historically, greenhouses were introduced for the overwintering of sensitive and exotic plants that would not survive the colder climates at northern latitudes, and temperature control remains perhaps the most important parameter to control in the greenhouse environment.

The control of temperature at night in the cooler seasons is achieved relatively easily with a well designed heating system. The high temperatures during the summer mean that the temperature inside will remain higher than that outside until the heat absorbed by the internal surfaces (benches, soil, plants, etc) is dissipated through radiation. The amount of latent heat within the greenhouse can be considerable as revealed in Table 5.2, which shows that a mature crop of tomatoes is a significant source of heat.

latent heat

Component	Latent heat (water equivalent Kg m^{-2})
soil	75 (depth 200 mm)
crop	10.2 (mature tomatoes)
heating system	1.7
glass	1.4

Table 5.2 Latent heat in greenhouse components (water equivalents, kg m^{-2}). Adapted from 'Crop Processes in Controlled Environments', Rees A.R, Cockshull K.E., Hand D.W. and Hurd R.G. eds, Academic Press, 1972.

∏   What consequences will the latent heat from the crop have for temperature control?

The latent heat of plants illustrates one of the fundamental features of green houses. They are dynamic systems and a variety of parameters change with time. For example, initially the contribution of the plants to the heat balance within the greenhouse is insignificant. As the crop matures, however, its heat content and heat generating capacity must be accounted for.

greenhouse effect

The control of day time temperatures is slightly more complex. Greenhouses heat up for two main reasons. Short wave radiation is converted to long wave radiation which is retained by the glass - the greenhouse effect - and there is limited exchange between the greenhouse and external atmospheres. It is essential, therefore, that cooling systems are used, and these are often supplemented with shading when outside temperatures are maximal.

So how are we to achieve the desired degree of control? The first thing to consider, of course, is whereabouts in the system to measure the temperature and how temperature measurement should be carried out.

The ideal is to measure the temperature of the plant directly. However, apart from being difficult and expensive, the plant surface will not have a uniform temperature, as shown in Figure 5.1. This is due to shading effects and the angle of leaves to the incident radiation.

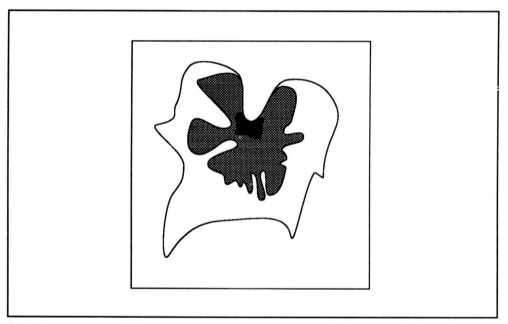

Figure 5.1 Temperature distribution on a cucumber leaf. Adapted from Bakker, J C, 'Acta Horticulturae' 148, 1984. Black = 33°C, grey = 31-33°C, white = 31°C.

The temperature of plants can exceed air temperature considerably at times. When transpiration is virtually eliminated leaf temperatures can be 28°C above ambient temperature. However, the discrepancy is usually not as excessive (Figure 5.2) so that the air temperature can be taken as an adequate and practical indicator of plant temperature. This proves satisfactory for most applications such as plant growth and experimental comparisons.

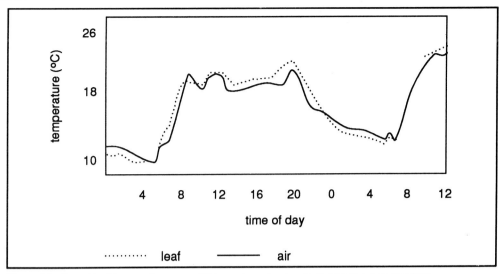

Figure 5.2 Air and leaf temperatures over 36 hours in a glasshouse in summer. Adapted from Burrage, S.W., 'Analysis of the Microclimate In Crop Processes in Controlled Environments', Acad Press, 1972, Eds Rees, Cockshull, Hand and Hurd.

5.2.1 Methods of measurement

The method chosen to measure the temperature depends upon the sensitivity and the response time required. A standard glass mercury thermometer will measure ±0.5°C if the change in temperature is fairly slow but due to the slow response times registers a change of only 0.25°C for an actual change of 1°C if temperature fluctuations are rapid. Thermocouples and thermistors have rapid response times and record almost 1°C for every 1°C change in temperature. Thermistors, or resistance thermometers, rely on the use of semi-conducting materials. These have a low conductivity which increases as the temperature rises. This change in resistance can be calibrated so that the temperature can be measured. Thermocouples are basically a loop of two conducting materials, for example copper and iron, with two junctions between the two metals. If both junctions are at the same temperature there is no current, however if one junction is warmer than the other then the resulting current can be used to measure the temperature. There may be a requirement for measuring rapidly fluctuating temperatures in a growth chamber under experimental conditions where the versatility of thermocouples in measuring soil, water, plant and air temperatures can also be exploited. In a greenhouse the response times will be relatively slow and the precision required is far less, so standard temperature measuring devices should be adequate.

thermocouples and thermistors

| SAQ 5.1 | What factors should be considered when deciding where to position measuring devices for temperature control in a greenhouse? |

5.2.2 Methods of heating

Heat can be supplied by a variety of systems each of which have their particular advantages and disadvantages. Typically, glasshouses are heated by radiator pipes or hot air ducts which may be supplemented with gas or electric heaters as the situation demands.

One disadvantage of the hot water radiator system is the response time. This is defined in terms of the time constant, which is the time taken for the system to reach 63% of its final value. The time constant is measured in hours for hot water pipes whereas hot air systems respond within minutes. However although the hot air system has the advantage of reacting quickly, there is little or no radiant heat produced. This means that high rates of heat supply are difficult unless the incoming air is very hot, which is clearly undesirable, or the cross section of the ducting is large. The best system utilises a combination of hot air and radiator pipes, often supplemented with localised sources to avoid cold spots.

Let us now examine how we might achieve a uniform temperature distribution throughout the plant growing zone of a standard 'A' frame glasshouse.

5.2.3 Heat distribution

The attainment of a uniform temperature throughout the plant growing zone is, perhaps, not as difficult as might first be envisaged. The horizontal and vertical temperature profiles obtained when steam coils are placed along the walls of the house are shown in Figure 5.3.

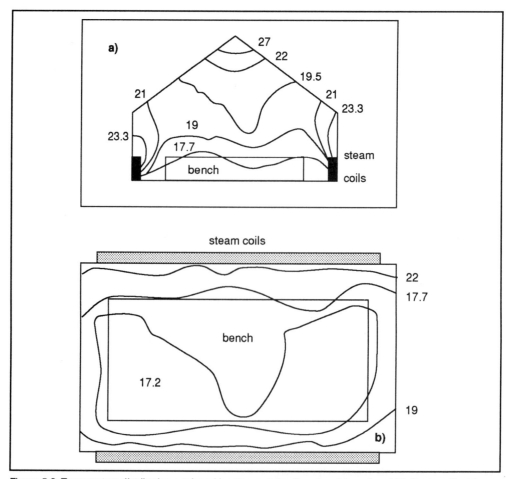

Figure 5.3 Temperature distribution produced by steam coils along the side walls. a) Vertical profile, b) horizontal profile 30 cms above bench. The lines are isotherms of the indicated temperatures. Adapted from Mastalerz, J W, 'Greenhouse Environment', 1977, J Wiley & Sons.

The warmest areas are close to the steam coils and in the roof space. The coolest place is in the centre under the bench, much as would be expected. Temperature gradients may be 6-8°C within 0.3 m of the wall but only 2-3°C over the rest of the growing zone as the system goes through the heating/cooling cycles. This degree of control can be improved by the use of a hot air circulating system which eliminates most of the residual temperature variation within the house.

SAQ 5.2

The greenhouse shown in Figure 5.3 has a centrally located bench but you probably know that many greenhouses have staging right against the glass. What would be the effect of this on plant growth and what might be done to rectify any problems?

5.2.4 Heat requirement

The amount of heat required to maintain the desired temperature within the greenhouse will depend on a number of factors. The most important of these is the nature of the material chosen to cover the greenhouse. The different insulating properties of the covering materials are expressed in terms of the thermal coefficient (K), which is the energy required to maintain the desired temperature per 1 m^2 of greenhouse per degree of temperature difference between the internal and external temperatures. The value of K will vary with wind speed and temperature and so is defined under standardised conditions of 0°C external temperature, 15°C internal temperature and 4 m s^{-1} wind speed. This value is then used to calculate the heat loss from a structure using the formula:

$$Hi = K A (T_i - T_o)$$ (Equation - 5.1)

where:

Hi = heat loss (W h^{-1})

K = thermal coefficient (W m^{-2} degree C^{-1})

A = area of covering (m^2)

T = temperature, inside (i) and outside (o) respectively.

An increase in the wind velocity will increase the heat requirement, by increasing heat loss through the covering, by a factor of approximately 1.1 at 6.7 m s^{-1}, and by a further factor of 0.04 for every extra 2.2 m s^{-1} above this. Wind will also affect infiltration; that is heat loss through any gaps in the structure. The heat loss through leaks can be calculated by considering the volume of air infiltrating and the capacity of air to remove heat (Equation 5.2). The amount of heat lost depends upon the temperature difference and the density of the air.

$$Hi = 3.04.V.\rho (T_i - T_o)$$ (Equation - 5.2)

where:

Hi = heat loss (W h^{-1})

V = volume of infiltrating air (m^3)

ρ = air density (kg m^{-3})

3.04 = specific heat of air (W kg^{-1})

These equations have been combined and their use simplified by the production of correction tables which take into account the various climatic and construction factors involved. An abridged version of these tables is presented in Table 5.3.

		Heat loss from	
		Gables (both) (W h^{-1})	Roof-length (W h^{-1} m^{-1})
Width of house	6.1 m	2344	1699.4
	11.0 m	7618	3047.2
		Heat loss from walls (W h^{-1})	
	Wall height	1.83 m	0.61 m
Wall length	6.1 m	2637	879
	11.0 m	4993	1664
	12.2 m	5567	1758
	15.2 m	7032	2344
	18.3 m	8204	2637

Construction factors (C)

glass = 1

Concrete block 10 cm thick = 0.58   20 cm thick = 0.46

(base on which frame is erected)

Structural factors (S)

metal frame clad with single glass and well sealed = 1.08

wood frame clad with single glass and poorly sealed = 1.25

wood frame clad with double glass and well sealed = 0.70

Climate factors for average wind speed and temperature conditions (K)

	Wind speed (m s^{-1})		
T_i-T_o °C	6.7	8.9	11.2
7	0.62	0.65	0.67
15.5	0.84	0.88	0.91

Table 5.3 Correction values for various climatic and construction factors for greenhouse heat loss calculations. Adapted from Nelson P V, 'Greenhouse Operation and Management', 3rd ed, 1985, Reston Publishing Co. (See text for discussion).

To calculate the heat requirement for a standard 'A' frame greenhouse clad with a single layer of glass, proceed as follows:

Find the heat loss values for the glass area (walls + roofs + gables) and the curtain wall (use the same table for glass and curtain wall). Multiply these values by the K factor for the appropriate temperature difference and wind speed. Then multiply by the construction factors for the structure and the material, adding the two figures to obtain the final answer.

Example 1

Calculate the heat requirement for a greenhouse of dimensions 6.1 x 15.2 x 1.83 m, with a curtain wall 0.61 m high (total height of house is therefore 2.44 m) made from 100 mm thick concrete. The house is clad with single glass, all metal frame, and is well sealed. Wind velocity is 6.7 m s^{-1} and the temperature difference T_i-T_o is 7°C.

Here, in outlinem is how the calculation is done.

Heat loss through the gable ends of a house 6.1 m wide is 2344 W h^{-1} + loss from the roof of length 15.2 m at a rate of loss of 1699.4 W h^{-1} for every metre is 25 830.88 W h^{-1} + loss through the transparent wall (glass) of total length 15.2 m x 2 (sides) plus 6.1 m x 2 (ends) therefore total = 42.6 m is 19 338 W h^{-1} and for the curtain wall 0.61 m high and 42.6 m in length is 6446 W h^{-1}.

The heat loss for the glass area is then 2344 + 25 830.88 + 19 338 = 47 512.88 W h^{-1}. This figure should then be corrected for climate, K = 0.62 at a wind speed of 6.7 m s^{-1} and a temperature difference of 7°C. For the construction material, glass has a C factor of 1.0. The house is well sealed, has a metal frame and is clad with single glass, the S factor is therefore 1.08.

Total heat loss from the glass area = 47 512.88 x 1.08 x 1.0 x 0.62 = 31 814.6 W h^{-1}.

The curtain wall figure is treated similarly, thus the construction factor C for 10 cm concrete block is 0.58 and the climate factor K is the same, 0.62. Structure does not apply in this case. Final value for the curtain wall is 6446 x 0.62 x 0.58 = 2317.98 W h^{-1}. The total heat loss for the entire house is therefore 31 814.6 + 2317.98 = 34 132.58 W h^{-1}.

We have drawn this series of calculations up as a kind of balance sheet in Panel 5.1.

Look through this panel carefully. When you think you have understood this balance sheet, then go on to attempt the SAQ.

SAQ 5.3

Calculate the heat requirement of a standard greenhouse of the following dimensions and specification. Width 11 m, length 21.3 m, height 2.44 m (including 0.61 m curtain wall). The house is of wood frame, double glass construction and is well sealed, curtain wall is composed of 20 cm thick concrete blocks. The wind speed is 8.9 m s^{-1} and T_i-T_o = 15.5°C.

Heat loss through the glass

Gable end	6.1 m wide	=	2344.0 W h^{-1}
Roof	15.2 m at 1699.4 W h^{-1} m^{-1}	=	25830.88 W h^{-1}
Walls	15.2 x 2 : 7032 x 2 = 14064		
	6.1 x 2 : 2637 x 2 = 5274		
	Total 19338	=	19338.0 W h^{-1}

Total = 47512.88 W h^{-1}

C factor = 1.0

S factor = 1.08

K factor = 0.62

∴ total loss through glass = 47512.88 x 1.0 x 1.08 x 0.62

= 31814.6 W h^{-1}

Heat loss through the curtain

Curtain is 0.61 m high

∴ 15.2 m x 2 : 2344 x 2 = 4688 W h^{-1}

6.1 m x 2 : 879 x 2 = 1758 W h^{-1}

Total = 6446 W h^{-1}

C factor = 0.58

S factor does not apply to a solid wall

K factor = 0.62

∴ total loss through curtain = 6446 x 0.58 x 0.62

= 2317.98 W h^{-1}

Grand total = 31814.6 + 2317.98 = 34132 W h^{-1}

Panel 5.1 Balance sheet for the intext activity.

5.2.5 Methods of cooling

Ventilation is the simplest method for cooling a greenhouse. The use of roof vents alone will cool the inside sufficiently provided that the outside temperature (T_o) is less than the inside temperature (T_i). The ventilation area is usually between 15 and 30% of the total floor area. This gives the greatest cooling to ventilation area ratio, beyond 30% there is only a small, 0.05°C, cooling effect for every extra % up to 45%. When the T_o is high the T_i can increase by as much as 11°C above ambient and further cooling is necessary. Forced ventilation with fans can keep the T_i to within 5°C of ambient which can be further reduced by the use of a fan and pad evaporative cooling system - Figure 5.4a.

cooling fan and pad

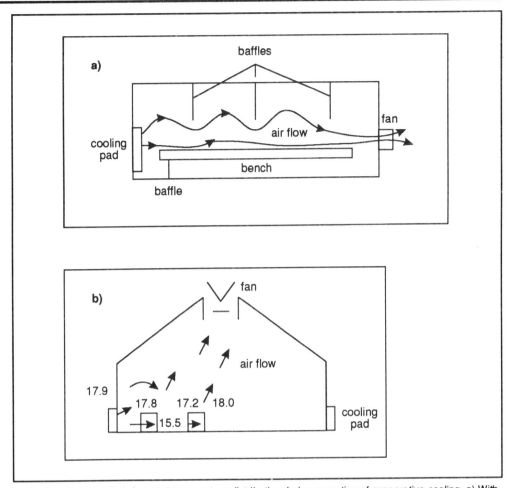

Figure 5.4 Patterns of air flow and temperature distribution during operation of evaporative cooling. a) With wall mounted fans, b) with ridge mounted fans. Only half the flow patterns are shown in b) for clarity. The right hand side is a mirror image of the left hand side. Adapted from Masatalerz, J W, 'The Greenhouse Environment', 1977, J Wiley & Sons.

The pad, which is open to the exterior, consists of cross fluted cellulose, or other suitable material which is built into the structure along one wall with extraction fans in the opposite wall. The pads are moistened with water which cools the outside air as it is drawn through the pad. If the external humidity is low this system can cool the incoming air by up to 14°C. Baffles are used to keep the flow of cool air over the plants. A slight temperature gradient is inevitable with this method of cooling and normally a differential of 4°C from pad to fan is accepted. This may be improved by positioning the fans in the ridge and drawing air through pads on both sides of the greenhouse (Figure 5.4 b).

A rate of air exchange of 2.5 m^3 min^{-1} m^{-2} of floor area is normally sufficient to maintain the desired temperature (for 300 m elevation and a light intensity of 53.8 klux, 968.4 μ mol m^{-2} s^{-1}). One square metre of pad should be used for every 45 m^3 of greenhouse volume and watered at a rate of 4.1 l min^{-1} for each linear metre of pad irrespective of the pad depth. Fans should have a maximum spacing of 7.6 m.

| SAQ 5.4 | Why do you think specifying the elevation and light irradiance are important when calculating rates of air flow? |

The capacity of the fans required in a cooling system can be conveniently calculated from tables that account for elevation (F elev), temperature (F temp), light irradiance (F light) and pad to fan distance (F vel) - see Table 5.4.

---

1) Elevation factor (F elev)

F elev = 1.04 at 300 m above sea level - add 0.04 for every 300 m extra up to 1500 m - below 300 m = 1.0

2) Temperature factor (F temp)

Desired variation in temperature between pad and fan	5.0°C	3.9°C	2.2°C
F temp	0.78	1.00	1.75

3) Light factor (F light)

F light = 1.0 for 53.8 klux - add or subtract 0.1 for every 5.4 klux difference

4) Velocity factor (F vel)

Distance from pad to fan	6.1 m	15.2 m	21.3 m	30.0 m
F vel	2.24	1.41	1.20	1.00

Table 5.4 Factors (abridged) for the calculation of required fan capacity when using evaporative cooling. Adapted from Nelson, P V, 'Greenhouse Operation and Management' 1985. Reston Publishing Co.

We will illustrate the use of these tables by using an example.

Example 2

Calculate the fan capacity needed for a greenhouse 15 m wide and 30 m long at an elevation of 900 m. A maximal light irradiance of 53.8 klux, 968.4 μ mol m^{-2} s^{-1} and a pad to fan temperature of 4°C is desired. The fan and pads are in the end walls.

The quantity of air to remove per minute is:

length x width x rate of air exchange = 30 x 15 x 2.5 = 1125 m^3 min^{-1}

Decide F elev, F tem and F light values and calculate their product to determine the overall value for the house (F house).

F elev is 1.12 for an altitude of 900 m, F temp and F light are both 1.00. Thus F house = 1.12 x 1.0 x 1.0 = 1.12.

The volume of air to be moved per minute should now be multiplied by the larger of the two values F house (= 1.12) or F vel (= 1.00 at 30 m)

The final value is therefore 1125 x 1.12 = 1260 m^3 min^{-1}.

Reducing the temperature of the incoming air is not a problem if a greenhouse overheats in winter. The main difficulty is to avoid temperature gradients which occur as a result of the colder heavier air falling to the floor of the greenhouse, forming a gradient as it gradually warms. Fan and tube ventilation systems are used to draw cold air in, mix this with the warm inside air and circulate it evenly through the tubes which run along the length of the house.

5.2.6 Heating economics

Heating a greenhouse is the most expensive outlay for the operator. Recent figures from Dutch nurseries put the cost of heating at 2-3 times that of providing supplementary lighting. Pressure from increasing oil prices and government policies aimed at reducing environmental pollution have focused attention on reducing fuel consumption and at the same time increasing profit.

5.2.7 Alternative coverings

Several plastic alternatives to glass have been evaluated for greenhouse cladding material. These have the advantage of a lower capital cost and now can be expected to have a reasonable lifespan. Some of these materials also have superior thermal properties to glass and are, therefore, attractive as energy saving materials (Table 5.5).

Cover	Saving % relative to glass	Light transmission (% of outside radiation)
single glass		76
double polyethylene	38	60
double polyvinylchloride	40	65
double polyvinylfluoride	34	59

Table 5.5 Energy saving with alternative cladding and the effect on light transmission. Adapted from Ferare J and Goldsberry K L, 'Acta Hort' 148, 1984.

They are also easy to use in double layers with an insulating air gap between and still remain relatively cheap compared to glass for which double glazing is prohibitively expensive for most applications. The use of double skin plastic coverings can save between 30 and 40% of the energy used to heat an equivalent glass house. However, there is a considerable reduction in the light reaching the crop which may have significant effects on production (see Section 5.3.4).

5.2.8 Thermal screens

thermal screens

The advantage of thermal screens is that they can be used at night, when heat loss is usually greatest, and retracted during the day to avoid light loss. Another approach is to use fixed screens during the season when heat loss is greatest, removing them completely for the rest of the year. This can again give energy savings of 40-50% with a loss of light radiation of typically 5%, although this has been reduced to 1-1.5% with improved equipment.

5.2.9 Effects of insulation on the greenhouse environment

We have already seen that there are conflicting requirements between the need to conserve heat and at the same time maximise the radiation available to the crop. Any method used to improve insulation is likely to reduce radiation and may also influence other aspects of the greenhouse environment, as shown in Table 5.6.

	Dec	Jan	Feb	Mar	Apr	May	Jun	Jul	Aug	Sep
Mean temperature (°C)										
Glass	19.8	19.8	20.4	20.3	20.2	19.7	20.7	20.5	21.2	20.1
Glass and thermal screen	20.2	20.5	20.4	20.4	20.3	19.9	20.7	20.3	21.3	20.2
Mean humidity (% RH)										
Glass	67	70	75	79	72	77	73	74	77	78
Glass and thermal screen	69	85	77	80	73	77	74	74	76	78

Table 5.6 Average air temperatures and humidities as influenced by the fitting of a permenant thermal screen. Adapted from Starkey, N G, 'Acta Hort', 174, 1985.

∏   What are the effects of insulation with a thermal screen on the temperature and humidity in a greenhouse (see Table 5.6)?

From the data in Table 5.6 it can be calculated that the average temperature in a glasshouse fitted with thermal screens (20.42°C) is slightly higher than in the glasshouse without insulation (20.26°C). There is a similar effect on the relative humidity where the average for the insulated house is 76.3% compared to 74.2% for the uninsulated house.

5.2.10 Plant response to changes in the temperature regime

The critical question, of course, is what effect does this have upon the plants? Table 5.7 gives the yield for tomatoes grown in the conditions described in Table 5.6.

	Feb	Mar	Apr	May	Jun	Jul	Aug	Sep	Total kg m^{-2}
Glass	3.7	3.7	5.0	6.1	6.0	5.8	5.5	4.3	40.1
Glass and thermal screen	3.9	3.9	4.9	5.6	5.6	5.8	5.4	4.3	39.3

Table 5.7 Yields of tomatoes under the conditions given in Table 5.6. Yields are reported as kg m^{-2} (data from Starkey, N G , Acta Hort 174/1985).

The total yield in the insulated glasshouse is slightly below that achieved in the uninsulated glasshouse, although in February and March the insulated house gave higher yields.

day/night temperature cycles

Alterations in the day/night cycle of temperature also affects the development of the plant, sometimes in unexpected ways. Figures 5.5 and 5.6 and Table 5.8 give data for tomatoes and sweet pepper grown under different day/night temperature cycles. A night temperature of 12°C increased the mean fruit weight in tomato at final harvest (13/6 Figure 5.5) considerably when compared to the 15°C treatment (Figure 5.5 a) but there was no financial gain from this as the production was delayed (Figure 5.5 b). The relative growth rate of sweet pepper was reduced when the night temperature was lowered quickly to 12°C but increased if the temperature change was slow (Figure 5.6). The lower night temperatures unfortunately give an unacceptably high level of mis-shapen fruit but this drawback can be overcome by increasing the day temperature to 25°C (Table 5.8).

Temperature (°C)		kg produced m⁻²	% misshapen
day	night		
20	20	4.2	33
20	15	4.5	24
25	15	5.1	2

Table 5.8 The effect of different day/night temperature regimes on the production and quality of sweet pepper. (Data from Kooistra, E, Acta Hort 148, 1984).

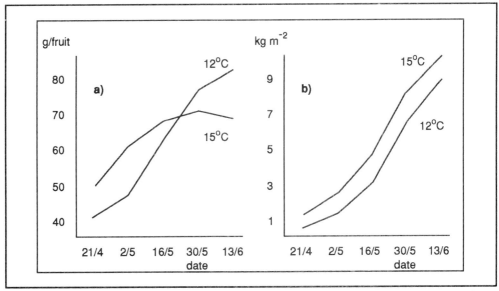

Figure 5.5 Average fruit weight a) and course of production b) of tomato with night temperatures of 12°C and 15°C. Adapted from Kooistra, E 'Acta Hort', 148, 1984.

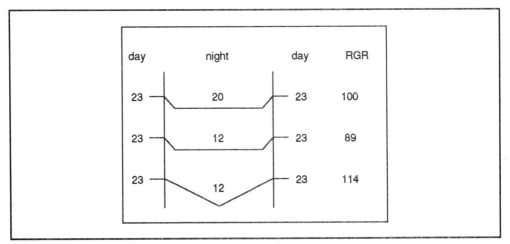

Figure 5.6 Effects of rate of change of night temperature (°C) on the relative growth rates (RGR) of sweet pepper. (Data from Kooistra, E. Acta Hort 148, 1984).

Π What conclusions can be drawn from the data above, regarding the response of
    plants to different temperature regimes? What might the practical consequences
    of this response be? (Try to formulate an answer to this before reading on).

5.2.11 Integration of temperature by plants

The data in Figures 5.5 and 5.6 and Table 5.8 indicate an important phenomenon in the
response of plants to temperature, ie the plant can compensate for poor growth and
development, induced by temperature changes in one phase of the day/night cycle, by
accelerated growth and development in response to counterbalancing temperatures in
the other phase. There is an 'averaging' or 'integrating' effect on growth. This is an
important observation for energy saving measures, as cooler night temperatures can
often be used with warmer day temperatures thus reducing energy costs. Production
can also be increased by manipulation of the temperature regimes. There is a dramatic
increase of over 25% in flower production if night temperatures of 24°C are used for
'Sonia' roses compared to the standard conditions (19°C). However the quality of the
flowers declines if the 24°C period exceeds 9 hours.

*effects of night
temperatures*

The conclusion to be drawn from these data is that we not only expect the temperature
optima for different species to vary considerably, but the manner in which the
temperature changes are achieved, ie gradually, stepped or abruptly, is also important.
There are several other important considerations when developing temperature
integrating regimes. Plant response will be different as the plant goes through different
developmental stages. Lower night temperatures are deleterious to cucumber before
the production phase, for example, but the response varies with variety. Of course some
species will not integrate temperature at all.

5.2.12 Root temperature

So far we have dealt only with the air temperature but you should be aware that root
zone warming can also influence yield. Root zone warming is achieved by the use of
heating cables placed in the soil or other growing media. A temperature of 33°C in the
root zone has been reported to result in a 9% increase in flower production in roses (cv
Royalty) compared to plants grown with roots at standard temperatures of 18°C.
Tomatoes respond well to root warming up to 24°C, after which yield and quality
decline. This response is seasonal, being more pronounced in the spring, when a 40%
increase can be achieved, than in the autumn where only a 7% increase is seen.

*root zone
warming*

5.2.13 Temperature control in growth chambers

The regulation of temperature within a growth chamber poses a slightly more difficult
problem than in the typical glasshouse in that a far greater degree of control is
demanded. A refrigeration system is used, balanced against heaters to ensure the
precise control needed.

*refrigeration
verses
heaters*

The basic layout of a modern growth chamber and typical temperature distribution
characteristics are shown in Figure 5.7 a and b. Air flow is vertical through the chamber
and the lights are housed in a separate compartment with its own cooling system.

Π How does the temperature control in the chamber shown in Figure 5.7 b compare
    with that shown in Figure 5.3?

Clearly, although there is a small vertical temperature gradient (<1°C) the growth chamber shows much less temperature variation than the greenhouse but, equally important, the chamber shows virtually no horizontal variation, which means that all plants are exposed to almost identical conditions.

SAQ 5.5

Examine the design of the growth chamber in Figure 7.7a. What is important about the design of the light box, ie parts A, B and C?

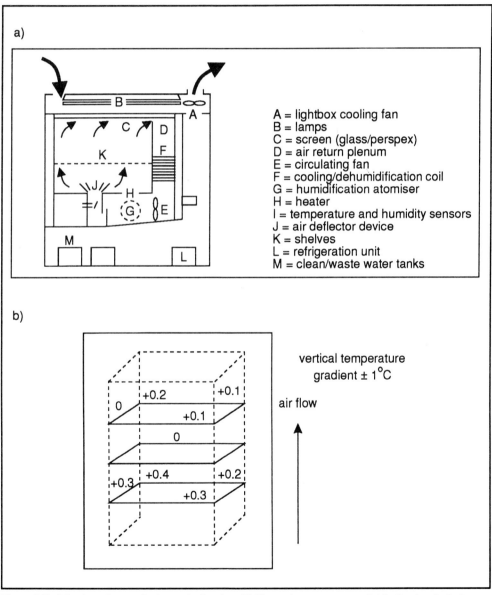

a)

A = lightbox cooling fan
B = lamps
C = screen (glass/perspex)
D = air return plenum
E = circulating fan
F = cooling/dehumidification coil
G = humidification atomiser
H = heater
I = temperature and humidity sensors
J = air deflector device
K = shelves
L = refrigeration unit
M = clean/waste water tanks

b)

vertical temperature gradient ± 1°C

air flow

Figure 5.7 a) Design of a standard growth chamber. b) Shows the temperature distribution in a typical growth chamber. The temperatures are recorded relative to the temperature at the centre of the chamber.

vertical air flow

We have already referred to the importance of attaining uniformity throughout the growing zone (Section 5.2.3) and a vertical flow is preferred for this reason. There will be a small but nevertheless significant temperature difference between the plants in the growth chamber if the flow of air is horizontal, whereas a vertical flow ensures that all the plants are exposed to the same gradient.

output varies with temperature

There is one other point to make about light and temperature. Fluorescent lamps are often used in growth chambers and their output varies with temperature (Figure 5.8). The optimum operating temperature for a lamp will be considerably different from the conditions chosen for plant growth and deviations from this optimum will reduce the output. Therefore, to ensure uniform illumination over a range of temperatures it is essential to regulate the temperature of the lights independently. There is a penalty to pay as the perspex screen typically reduces the available light by about 4%. This is offset by the dramatic reduction in conducted and convected heat exchange, reduced wear of the lights (corrosive effects of high humidity etc are avoided) and the lessened possibility of contamination.

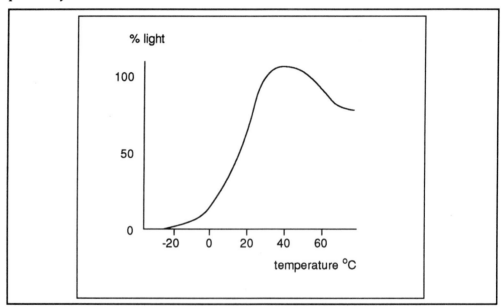

Figure 5.8 Light output of a fluorescent lamp as a function of temperature. Adapted from Downs, R J and Hellmer, H, 'Environment and the Experimental Control of Plant Growth', 1975, Academic Press.

5.2.14 Cooling capacity

The refrigeration unit must have an adequate capacity to ensure that good temperature control is achieved. There are a number of different heat sources that contribute to the overall cooling requirement in a growth chamber that must be considered when calculating the heat load.

These include the heat from the lights (radiant, conducted and convected), transmission through the walls of the cabinet, difference in temperature between the exchange air (from outside the chamber) and the set temperature within the chamber, plus the latent heat loads due to water vapour as well as a number of factors that are difficult to quantify such as the plants, soil, pot sizes, spillages etc. These are beyond the scope of this book but details can be found in 'Controlled Environments for Plant Research', R J Downs, 1975, published by Columbia University Press, New York and London.

5.3 Light

5.3.1 Photochemical reactions

The utilisation of light to fix atmospheric carbon in the photosynthetic process is the principal photochemical reaction in plants. Optimising the level of irradiance to achieve the maximum rate of photosynthesis and hence crop production is the main objective in the greenhouse. There are, however, a number of other responses of plants to light that are vital for regulation of growth and development (Figure 5.9).

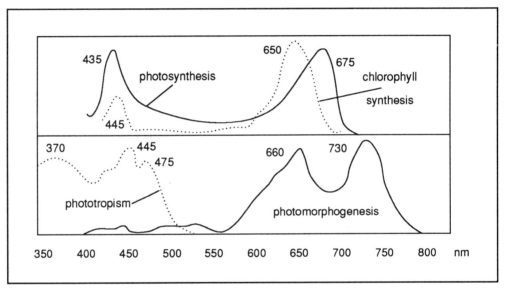

Figure 5.9 The photochemical action spectrum of plants. Adapted from 'Lighting for Plant Growth', Bickford, E.D. and Dunn, S., 1972, Kent State University Press.

four type light reactions

There are four principal photochemical reactions (Figure 5.9) which we are concerned with:

- chlorophyll synthesis is initiated by light from the wavelengths between 350-470 and 570-670 nm with optima at 445 nm and 650 nm;

- photosynthesis utilises light most effectively between 350-530 and 600-700 nm;

- phototropism, which controls plant growth toward or away from light, responds to light between the wavelengths 350-500 nm with optima at 370, 445 and 475 nm;

- photomorphogenesis through the red and far red phytochrome receptor system which is activated by wavelengths from 570-800 and 680-780 nm with optima at 660 and 730 nm.

5.3.2 Light measurement

When constructing protected environmental facilities it is obviously desirable to know the irradiance and spectral distribution of light from a natural or an artificial light source.

Figure 5.10 shows the wavelengths that can be measured by the different types of available sensors and also shows that the sensors are more sensitive to some wavelengths than others.

We will not discuss the physical processes involved in these types of sensors but we will examine the suitability for use in the measurement of light in relation to the design and operation of greenhouses.

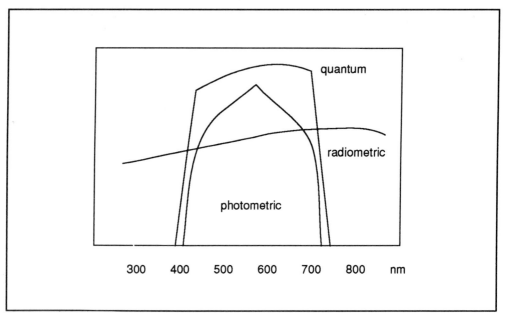

Figure 5.10 Relative sensitivities of different types of photosensor used in the measurement of 'light' (radiant energy).

∏ Study Figures 5.9 and 5.10 and determine which type of measurement will be the most appropriate for deciding the most suitable light source to support plant growth and development.

You can see that the various methods of measurement either do not cover the entire plant response spectrum or have a 'flat' response. The correct answer, therefore, is that none of the methods shown are entirely appropriate for the following reasons.

Photometric method - (expressed in lux) is designed to reflect the response of the human eye.

Radiometric method - (expressed in $W\,m^{-2}$) gives a measure of the total energy. The instruments used are equally responsive at all wavelengths and are therefore only relevant to plant response in a rather indirect way. For instance sunlight and incandescent lamps give unusually high readings compared to other methods, because of the large emissions at long wavelengths (780 nm).

PAR Quantum method - (expressed in $\mu mol\,m^{-2}\,s^{-1}$) gives a measure of the photosynthetically active radiation (PAR) available. This is the method of choice as the wavelengths covered correspond to the action spectrum for photosynthesis and most accurately reflect the energy available for growth. However there is no spectral sensitivity; response is again virtually equal at all wavelength between 400-700 nm.

Crop production correlates reasonably well with PAR under fluorescent sources (Table 5.9) but there are discrepancies. These differences are greater when plants are grown under a mixture of fluorescent and incandescent lamps or natural sunlight.

Lamp	Plant growth (mg dry mass/unit of PAR)
Phillips 33	3.5
Phillips 55	3.0
Phillips 37	3.0
Phillips 32	3.7
Phillips 83	3.3
Phillips 84	3.1
Phillips 34	3.0
GEC 32	3.7
GEC 29	2.8
GEC 50	3.3

Table 5.9 The growth response of plants grown under different fluorescent lights. The results are expressed as mg of dry mass per unit of PAR. Adapted from Andersen, A, 'Acta Horticulturae', 174, 1985.

The main reason for this is that the part of the spectrum from 710-780 nm which elicits photomorphogenic responses is not accounted for with PAR measurements, and there are other minor but nevertheless significant effects on photosynthesis of wavelengths outside the 380-710 nm band.

From Figure 5.9 we can see that the action spectrum for plants extends from about 350 to 780 nm and that there is a greater sensitivity to certain wavelengths.

In practice all three sensors are commonly used and photometric or radiometric data can be approximated to PAR by conversion factors (Table 5.10).

∏ Why are the conversion factors shown in Table 5.10 different for different lamps?

Since the emission spectra are different for different types of lamps, the amount of the light emitted by these lamps which will be detected depends upon the absorption spectra of the sensor. This is different for different sensors. Thus, in converting the amount of light detected to the amount of PAR, we should anticipate different conversion factors for the two sensors and for different lamps.

Initial measurement	Convert to PAR	Daylight	Metal Halide	Sodium	Mercury	White Fluorescent	Tungsten
radiometric measurement (W m^{-2})	μmol m^{-2} s^{-1}	x 4.6	x 4.6	x 5.0	x 4.7	x 4.6	x 5.0
photometric measurement (klux)	μmol m^{-2} s^{-1}	x 18	x 14	x 14	x 14	x 12	x 20

Table 5.10 Approximate conversion factors for converting photometric and radiometric data to PAR.

5.3.3 Artificial sources of light

The typical growth response of plants to single artificial light sources is summarised in Table 5.11. Although growth occurs under all light sources, plants may show different morphological and biochemical features compared to plants grown in daylight.

Fluorescent cool white (FCW 9.38, 0.18): green foliage, stems elongate slowly, multiple side shoots develop, flowering occurs over a long period of time. High intensity discharge mercury (HG 55, 3.6) and metal halide (MH 122, 14.8) lamps give a similar response to fluorescent lamps at equal energy.

Fluorescent Gro-lux (FGL 5.86, 0.07): deep green foliage often larger than with FCW, stems elongate very slowly, multiple side shoots develop, flowering occurs late and flower stalks do not elongate. The effect of high pressure sodium (HPS 123, 46.5) lamps are similar at equal energy.

Incandescent (INC 6.9, 8.11): foilage pale and leaves are longer than with other sources, excessive stem elongation, spindly, side shoots suppressed, flowering rapid, plants mature quickly and senesce.

Table 5.11 Generalised plant growth responses to single artificial light sources. Adapted from 'Light and Lighting Systems for Horticultural Plants', Cathey, H M and Campbell, L E, Chapter 10, Horticultural Reviews 2, 1980. Figures in brackets give the radiation output in watts at 400-700 nm and 700-850 nm respectively.

Figure 5.11 gives the spectral energy distribution of several different light sources.

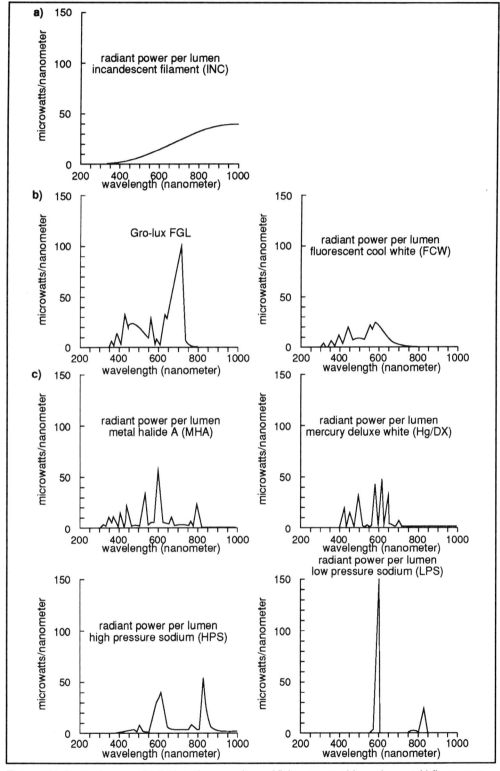

Figure 5.11 Spectral energy distribution of commonly used light sources. a) Incandescent, b) fluorescent, c) high intensity discharge.

∏ Consider Figure 5.11. From this and the information given in Table 5.11, what do you think will be the best artificial light source to support plant growth?

You will notice that, with minor variations, all fluorescent sources are deficient in longer wavelength radiation, 650-780 nm, whilst incandescent (tungsten) lamps emit at these and longer wavelengths. So although fluorescent lighting alone will support growth, a combination of fluorescent and incandescent sources gives a considerable improvement (Table 5.12).

	Fluorescent (gro-lux) and incandescent	Fluorescent (gro-lux)
number of leaves	40	32
fresh wt (shoot) g	103	89
number of flowers	127	90

Table 5.12 Growth and flowering of African violet under different light sources at 25°C, 16 hour day length and equal energy input. Adapted from 'Lighting for Plant Growth', Bickford, E D and Dunn, S, 1972, Kent State University Press.

combining fluorescent and incandescent light sources

The amount of energy contributed by the incandescent lamps, which have a red to far-red (R/FR) ratio of about 0.75:1, should be approximately 10% of the total energy supplied in order to reduce the R/FR ratio (which is about 5:1 for fluorescent sources) to 2.3:1. This difference in R/FR is largely responsible for producing the abnormal and opposite growth responses when single light sources are used.

The precise ratio of R/FR used and the timing of illumination will depend upon the plant species and the desired effect on development.

SAQ 5.6

Describe the growth response of a plant grown under low pressure sodium lamps (LPS 63, 8.6) and no other light source, explaining your answer. Consult Figure 5.11 and Table 5.11 again.

5.3.4 The relationship between light and crop production

relationship between radiation received and yield

The productivity of greenhouse crops is heavily influenced by the amount of light they receive. In order to optimise the conditions for production within the greenhouse it is necessary to know the precise relationship between incident radiation and yield. Data for tomato (Figure 5.12) show that yield is accumulated in direct proportion to solar radiation received (other factors held constant). This means that for every 1% reduction in light there is a 1% loss in yield. Therefore any measures that reduce the amount of light, such as screens etc, to reduce energy costs, must be balanced against the loss of production. Carry out the following calculation which will help to illustrate this point.

SAQ 5.7

Assuming a return of £25 m^{-2} for a crop of tomatoes and a fuel cost of £4 m^{-2}, what percentage fuel saving is required, to make cost effective, an energy saving measure that reduces incident radiation by 2%?

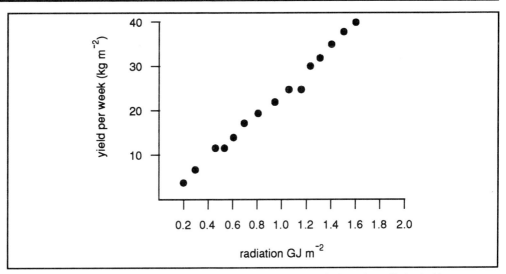

Figure 5.12 The relationship between cumulative total yield and cumulative solar radiation incident on a crop of tomatoes. Adapted from Cockshull, K E, *et al*, J Hort Sci, 68, 1992.

A reduction in the amount of light reaching a crop does not only affect the yield. The marketable quality of the fruit may also be reduced in terms of size classes as shown by the data in Table 5.13. Flower abortion may occur under heavy shading.

% shade	Fruit size (% each class)			
	>57 mm	52-57 mm	47-52 mm	<47 mm
0	13.4	52.5	25.5	8.7
6.4	9.6	50.8	28.9	10.9
23.4	5.8	44.1	34.5	15.8

Table 5.13 The effect of shading (percentage loss of available light) on fruit size in tomatoes. Data from Cockshull, K E, *et al*, J Hort Sci, 68, 1992.

excess light — Over-exposure to light can also be deleterious to growth and result in plant injury. Chlorosis and reduced leaf growth occur after 5-7 days of continuous light in tomatoes and interestingly enough the phenomenon is not related to total irradiance but to the duration of exposure.

Excess light is not a common problem, the major concern being to avoid the consequences of light deficiency, ie reduced photosynthesis, low carbohydrate content, thin epidermis and cuticle, reduced growth rate, delayed flower induction, increased risk of fungal infection.

5.3.5 Lighting installation

A uniform light irradiance is necessary throughout the growing region or uneven growth responses will occur as we have seen with temperature. Fluorescent lights provide a diffuse source that gives fairly uniform light over a wide area. Incandescent lights, however, emit between 80 and 90% of their light at infra-red wavelengths.

∏   What consequences does this have for greenhouse lighting?

Incandescent sources produce far more heat than fluorescent lamps. Incandescent sources are point sources, that is there is little diffusion of the emitted light. Consequently, large numbers of small lamps fitted with reflectors should be used to avoid localised heating. Increasing the number of light fittings will shade some of the natural light and therefore a balance must be reached to maximise the benefit. This has become less of a problem since high intensity discharge lamps have become available, producing equivalent light from far fewer sources.

5.3.6 Practical application: supplemental lighting, photoresponse control

supplementary lighting

Perhaps the most important use of lighting in horticulture is to supplement that which is available naturally. This need not be at high irradiance levels and it is not necessary to reproduce the full 'natural' daylight spectrum. Supplemental lighting is used to extend the day length, either pre-sunrise, post-sunrise or both and to provide lighting during all or part of the night.

African violets have improved colour and flower profusely when fluorescent lights are used at about 960 μmol m^{-2} s^{-1} PAR. Azaleas are forced to flower with continuous night lighting and the rate of forcing increases with the energy supplied. Most types of light elicit this response but incandescent sources produce a particularly good response. Orchids respond best to Gro-lux wide spectrum lamps at around 960 μmol m^{-2} s^{-1} PAR.

∏ What do these effects tell you about the use of artificial lighting in horticulture?

This question has been asked to emphasise the point that the type, intensity, duration and timing of illumination will depend upon the plants, not forgetting varietal differences, and of course the desired end product.

5.3.7 Plant adaptability

light compensation point

Plants grown in protected environments under sub-optimal light regimes have the ability to acclimatise to these lower light levels. Figure 5.13 reveals that this acclimatisation involves a reduction in the light compensation point (LCP), that is the point at which CO_2 fixation through photosynthesis balances the loss through respiration.

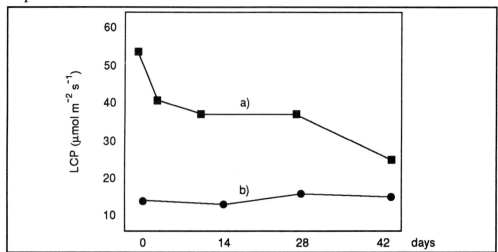

Figure 5.13 The effect of irradiance level on the LCP of Begonia. a) Plants transferred from 135 to 15 μmol m^{-2} s^{-1}. b) Plants kept at 15 μmol m^{-2} s^{-1}. LCP = light compensation point. Adapted from Fjeld, T, Gartenbauwissenschaft 57, 1992.

This reduction in the LCP is due mainly to a lowering of the respiratory rate which falls by approximately 50%, whilst photosynthesis is reduced by about 15%.

SAQ 5.8	When solar radiation is particularly low as in winter, the temperatures within the glasshouse environment may be considerably above ambient. An increase in temperature from 15°C to 30°C has little effect on the rate of photosynthesis at atmospheric CO_2 concentrations, it will increase by about 25% over this range, but the Q_{10} for respiration is about 2, that is it will double for every 10°C rise in temperature.
	What consequences could this have for the light compensation point and hence the plants, and how can this be remedied?

5.3.8 Lighting for growth chambers

The light for growth chambers is supplied entirely from artificial sources, consequently a number of other factors must be considered.

∏ Examine Figure 5.14 carefully. What does it tell you about the distribution of light in a growth chamber?

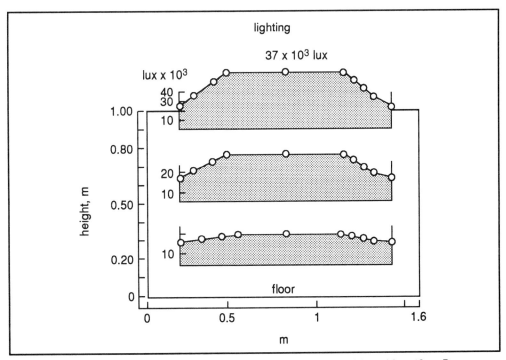

Figure 5.14 Light distribution in a growth chamber illuminated from above. Adapted from 'Crop Processes in Controlled Environments' eds Rees, Cockshull, Hand and Hurd, Acad Press 1972.

There is not a uniform distribution of light. The intensity declines sharply near the walls and away from the lights. This can be remedied by using lights along the walls, or more usually by increasing the reflectivity of the wall surface by covering it with aluminium.

Maintaining the desired level of light is also more of a problem in a growth chamber. The output from the lamps will decline with age and, therefore, must be monitored and spent lamps must be replaced as required. Recent developments in lamp design, such as the Phillips high frequency fluorescent tubes, can avoid this problem as they are dimmable which allows the exact light level to be set and maintained. A further point to note about these lamps is that the intensity can be programmed to increase gradually, avoiding the abrupt change from dark to full light and more closely simulating the natural situation.

dimmable lamps

5.4 The gaseous environment

5.4.1 Carbon dioxide

Plants fix CO_2 through photosynthesis and release it through respiration. Normally there is a net gain of CO_2, and hence growth, at normal atmospheric CO_2 concentrations of about 330 µl l^{-1} (ppm). There is a CO_2 concentration, however, at which photosynthetic gain is balanced by respiratory loss, the compensation point. This is about 50 µl l^{-1} for C3 plants such as bean and as low as 2-5 µl l^{-1} for C4 plants such as maize (Figure 5.15).

CO_2 compensation point

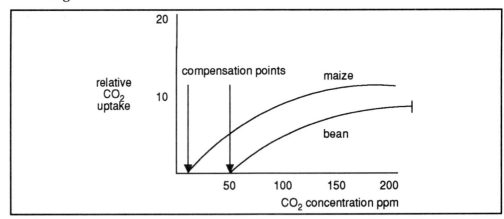

Figure 5.15 Effects of [CO_2] on CO_2 uptake in C3 (bean) and C4 (maize) plants. Adapted from Bidwell, R G S, 'Plant Physiology', 1979.

∏ See if you can think what the consequences of this are for the CO_2 levels in an enclosed environment where plants are growing?

It is evident that there will be a reduction in the CO_2 level as the plants continue to accumulate CO_2 until the compensation point is reached, after which there will be no net photosynthetic gain. The reduction in CO_2 can be considerable and rapid. Actively growing maize can deplete CO_2 from ambient levels to 80 µl l^{-1} within 40 minutes in the confined space of a growth chamber. The reductions caused by crops in greenhouses are not as dramatic but are nevertheless significant (Figure 5.16).

∏ The reduction in CO_2 by the carnation crop is likely to have an undesirable effect. What do you think this will be?

You can see quite clearly from Figures 5.15 and 5.16 that the rate of CO_2 assimilation decreases steadily as the CO_2 is depleted, so the carnations in the CO_2 depleted greenhouse, where the CO_2 levels are lowered to below 200 µl l^{-1}, will be growing at less than optimal conditions. The obvious remedy for this is to increase the CO_2 level by opening the vents in the greenhouse and ensuring that there is sufficient air exchange to meet the CO_2 demand of the crop.

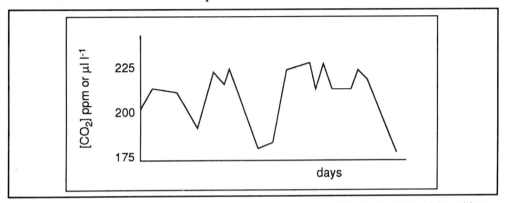

Figure 5.16 CO_2 depletion by a carnation crop during winter in an unventilated greenhouse. Adapted from 'Carbon Dioxide Enrichment of Greenhouse Crops', Volume 2, Enoch, H Z, CRC Press, 1986.

∏ What problems will this pose for maintaining a controlled environment within the greenhouse?

We are back to the problem that altering one factor inevitably results in changes in another. The opening of vents is likely to cause significant changes in temperature. Opening the vents and increasing air flow may well be a possibility in the summer, but the fall in temperature in the winter would be unacceptable.

5.4.2 Carbon dioxide enrichment

The solution to the problem just described in the previous section may be to enrich the atmosphere within the greenhouse with CO_2. This has the further benefit that the CO_2 concentration can be enriched far above the normal ambient level. Figure 5.17 shows data which explain the nature of the benefits of using this approach. The increase in the rate of CO_2 fixation by photosynthesis as the external CO_2 concentration rises is dramatic and has not reached saturation at $1400\,\mu l\,l^{-1}$, more than four times the ambient CO_2 level.

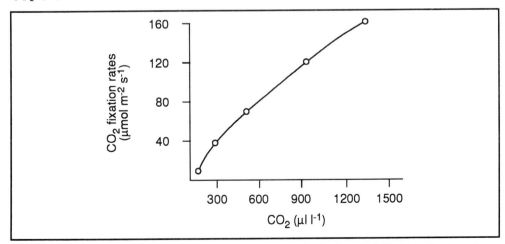

Figure 5.17 The effect of CO_2 concentration on the rate of CO_2 fixation in typical C3 plants. Re-drawn from Enoch and Kimball in, 'Carbon Dioxide Enrichment of Greenhouse Crops' Vol 2, Ed Enoch, CRC Press,

enhanced
yield
It is reasonable to expect from such a large effect on net photosynthesis that the response in terms of yield and production rates would be considerable. The effect of CO_2 on some typical greenhouse crops is given in Table 5.14. This shows that the effects are extremely beneficial. It also very often improves quality. However there are few instances where supplementing the CO_2 level does not improve yield.

Crop	Effect
Tomatoes	enhanced growth rate, more fruit per truss, 25% increase in yield
Lettuce	enhanced growth rate, harvest 2 weeks earlier, better quality
Roses	improved quality, 20% more yield
Chrysanthemum	faster growth rate, improved quality

Table 5.14 The effect of CO_2 enrichment to between 800 µl l^{-1} and 1000 µl l^{-1} on some greenhouse crops. Data from Enoch and Kimball. 'Carbon Dioxide Enrichment of Greenhouse Crops'. Vol 2, Ed Enoch, CRC Press 1986.

5.4.3 Methods of enrichment

The source used to enrich CO_2 levels ranges from compressed CO_2 cylinders, through release from carbonate salts by the action of acid, to the use of sewage sludge and industrial waste. The most common and convenient method for supplying CO_2 is from the flue gases of the heating system and this has the extra advantage of not incurring additional costs. The gases should be distributed evenly throughout the house via ducts and lateral pipes and are usually controlled to give a maximum level of 1000 µl l^{-1}. This concentration may not be optimal for all plants but can be achieved readily in practice and avoids some of the injuries that can occur with sensitive plants.

5.4.4 Light and carbon dioxide enrichment

light saturation   The radiant energy required to saturate photosynthesis at normal carbon dioxide concentrations of about 300 µl l^{-1} is 460 µmol m^{-2} s^{-1} (PAR). As the concentration of CO_2 increases so the amount of radiation needed to saturate photosynthesis increases (Table 5.15). These results are taken using a single leaf, the light levels required to saturate a full canopy will be much higher. Similar results are shown in Figure 5.18.

CO_2 concentration (µl l^{-1})	Saturating energy (µmol m^{-2} s^{-1})
300	460
500	676
1500	920
2000	993

Table 5.15 The relationship between carbon dioxide concentrations and the radiant energy required to saturate photosynthesis.

∏   What conclusion can you draw from the data in Table 5.15 about the light intensities at which CO_2 enhancement should be used?

From these results it would be reasonable to assume that it would be appropriate to use CO_2 enrichment at high light intensities. However, there are other factors to consider.

∏ Examine Figure 5.18 and Tables 5.15 and 5.16 and decide whether it is worth enriching with CO_2 at all PAR levels.

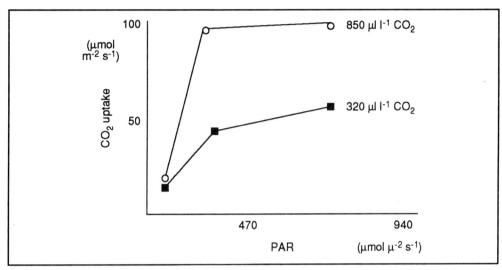

Figure 5.18 The influence of light intensity on the photosynthetic rate of a tomato leaf at different CO_2 concentrations. Adapted from Enoch and Kinball, 'Carbon Dioxide Enrichment of Greenhouse Crops'. Vol 2, Ed Enoch, CRC Press 1986.

PAR μmol m^{-2} s^{-1}	CO_2 concentrations (μl l^{-1})			Increase %
	330	1000	1600	
44	34	39	38	15
129	57	65	69	21
268	76	88	88	16
395	86	97	98	14

Table 5.16 The effect of CO_2 concentration on the mean relative growth rate (mg g^{-1} dry weight day^{-1}) at different PAR (μmol m^{-2} s^{-1}) in Chrysanthemum. The % increase is a comparison of the data obtained using 1600 μl CO_2 l^{-1} and that obtained using 330 μl CO_2 l^{-1}.

These data show that CO_2 enhancement is extremely effective even at low light intensities and other studies show that plant quality in roses, *Saintpaulia* and *Begonia* amongst others, is improved. It can also be seen that there is a proportionately greater effect of CO_2 at low light intensities: see Figure 5.18. These observations, allied to the problems with transpiration that occur at high light intensities, mean that CO_2 is most effectively used at lower light intensities.

5.4.5 Pollutants

Combustion of fossil fuels produces a number of other compounds apart from CO_2, most of which are potentially injurious to plant growth. Carbon monoxide (CO), nitrous oxides (NO, NO_2), ozone (O_3), sulphur dioxide (SO_2) and ethylene (C_2H_4) are among the compounds produced. The quantities of these gases produced depend upon the quality of the fuel used and the condition of the burners. Needless to say the burners should be well maintained and optimally tuned. Safeguards are also essential to close down the enrichment system should the burner fail, the gas mix temperature becomes too high or there is incomplete combustion.

The levels of pollutants that can cause injury are in some cases extremely low and photosynthesis can be affected by nitrous oxides that cause no visible signs of injury. NO_2 and O_3 at 50 ppb and SO_2 at 25 ppb (parts per billion) will damage tobacco plants within a few days and C_2H_4 can affect plants at concentrations of 5 ppb. The importance of selecting the fuel source and controlling pollutant levels can be appreciated from Figure 5.19. This shows the level of pollutants in a CO_2 enriched glasshouse in Holland where the natural gas used is a particularly 'clean' source.

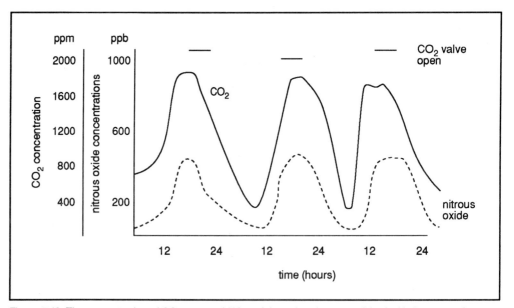

Figure 5.19 The concentration of CO_2 ppm and NO_2 ppb in a glass house enriched with flue gases from a natural gas burner, over three days in January with closed ventilators and thermal screen. Adapted from Wolting, H G, *et al* 'Acta Hort', 174, 1985.

It can be seen that under these conditions the pollutant level becomes sufficient at times to cause serious problems, so that although the enhanced CO_2 level has a protecting effect against NO_2 stress, the beneficial action of elevated CO_2 levels will be partly neutralised.

5.4.6 Plasticisers

Coverings such as polyvinylchloride (PVC), are sometimes used as a cheap alternative to glass for many horticultural applications. The flexibility required in PVC is achieved by incorporating alkyl esters of phthalic acid. These plasticisers are used in quantities of up to 50% by weight and act by weakening the polymer-polymer bonds in the PVC.

dibutyl
phthalate

There have been several instances of poor plant growth and even death associated with the increased use of plastics in the glasshouse industry. These problems have often been traced to the accumulation of dibutyl phthalates in the glasshouse atmosphere (Figure 5.20).

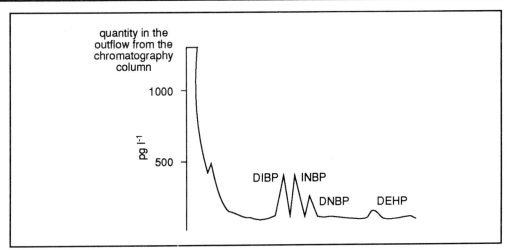

Figure 5.20 Gas chromatography trace of samples taken from the atmosphere in a glasshouse where plastics have been used to cushion the glass against the metal glazing bars. Adapted from Hardwick, r c, *et al* 'Ann Appl Biol', 105, 1984. Diisobutylphthalate (DIBP), iso n-butyl phthalate (INBP), di-n-butyl phthalate (DNBP), di ethyl-hexyl phthalate (DEHP).

These volatile compounds are easily lost into the atmosphere, particularly under greenhouse conditions where the temperature is warm in comparison to the surroundings. They are present in considerable quantities, certainly at levels high enough to cause damage to crops (Table 5.17).

Month	Temp (°C)	% time vents open	DBP (pg l⁻¹)
July	23.1	88.7	0920
August	22.1	79.6	1027
September	17.7	42	1020
October	17.6	34	1395

Table 5.17 Total dibutyl phthalate concentrations detected in the greenhouse atmosphere in pg l⁻¹. Data from Hardwick *et al* Ann Appl Biol 105, 1908.

Members of the *Brassica* family are particularly sensitive and *Brassica oleracea* (var Derby day) exhibit severely restricted growth, chlorosis and cotyledon death at 900 pg l⁻¹ and are only slightly less affected at 360 pg l⁻¹.

A further aspect of the use of plastics in the construction of facilities for plant growth is that they are susceptible to photodegradation leading to the release of other, possibly more volatile and toxic products.

This example serves to illustrate the need to test materials used in the construction of environmental facilities for phytotoxicity and to make you aware that materials proving negative in the initial tests for phytotoxicity may well release damaging compounds as they undergo degradation.

5.5 Humidity

5.5.1 Absolute and relative humidity

The mass of water held in the atmosphere is its absolute humidity and is expressed in grams per cubic metre ($g\ m^{-3}$). The absolute humidity will remain constant if no more water is introduced to the atmosphere. However, the ability of the atmosphere to hold water increases as the temperature increases, ie the saturated vapour pressure of the air rises. Thus the term relative humidity is often used. Relative humidity (RH) is the relationship of the actual water vapour pressure to the saturated water vapour pressure at a specified temperature, ie the ratio of the amount of water vapour present to the maximum amount of water vapour that could be held at that temperature. It is expressed as a percentage:

temperature affects on relative humidity

$$\%\ RH = \frac{\text{partial vapour pressure due to water}}{\text{saturated vapour pressure due to water}}\ \times\ 100 \qquad \text{(Equation - 5.3)}$$

at a specified temperature.

∏ Use Figure 5.21 to predict what will happen to the absolute and relative humidity as the temperature decreases from 20 to 16°C?

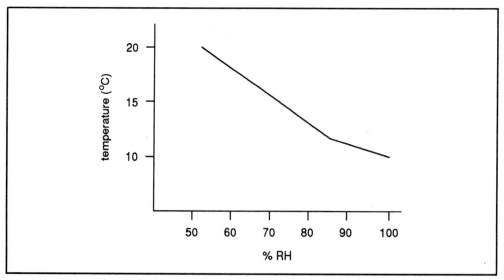

Figure 5.21 The relationship between relative humidity and temperature at a fixed absolute humidity of 10 g m^{-3}.

As long as no more water is introduced, the absolute humidity does not change, but the relative humidity increases as the temperature falls, so that at 20°C RH = 53%, and at 16°C RH = 66%.

∏ What is the relative humidity at 10°C for the system depicted in Figure 5.21?

From Figure 5.21 it can be seen that the relative humidity at 10°C is 100% or saturation. This is the dew point and refers to the temperature at which water will start to condense out.

5.5.2 Dew point

As described in the previous section, the amount of water vapour in the air can be expressed in either absolute or relative terms. However, as we have just seen, the water content can remain constant in absolute terms as the temperature is reduced until, at the dew point the actual vapour pressure it exerts is equal to the saturated vapour pressure. At temperatures below the dew point, water will condense out.

5.5.3 Measurement of humidity

Hygrometers

The most common form of hygrometer makes use of the fact that hair will expand and contract with changes in water content. The instruments based on this principle have low sensitivity, a long lag time and, therefore, do not detect sudden fluctuations. They have a precision of 6% at best. Reliability is generally good, however, and sensitivity can be improved by placing the unit in an aspirated chamber.

Electrical hygrometers rely on the changes in resistance of hygroscopic salts, such as lithium chloride, as the humidity changes. These devices have an error of less than 1.5% but have the disadvantage of failing if water condenses on them.

Psychrometer

The psychrometer consists of a wet and dry bulb thermometer. The evaporation of water from the water-soaked cloth covering the bulb cools the wet thermometer and the degree of cooling depends on the relation between the humidity and the rate of evaporation. The accuracy of the measurement is improved if air is drawn over the thermometers at a constant velocity of about 3 metres per second. Using the wet and dry bulb temperatures the humidity can be determined from standard tables and charts (Figure 5.22).

∏ Let us have a go at reading the chart shown in Figure 5.22. What is the relative humidity if the wet bulb registers a temperature of 20°C and the dry bulb a temperature of 30°C?

The answer is approximately 45%. To find this value, follow the diagonal line (sloping left to right) from 20°C on the wet bulb temperature axis and draw a line vertically from 30°C on the dry bulb temperature axis. Mark where these two meet and starting from this point, follow the curved lines to the relative humidity axis.

∏ Now determine the relative humidity when the wet bulb registers 30°C and the dry bulb also registers 30°C.

The answer is 100%.

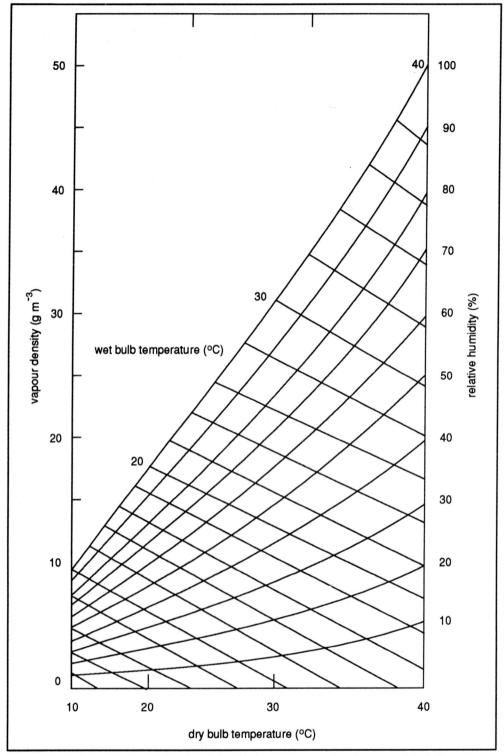

Figure 5.22 Psychometric chart giving the relationship between vapour density, wet and dry bulb temperatures, and relative humidity.

5.5.4 Control of humidity

Humidity is perhaps the most difficult aspect of the environment to control within a chamber or greenhouse. Clearly, to increase humidity it is necessary to increase the amount of water vapour in the atmosphere. This is done by using mist, evaporation from the wet pads used to cool in temperature control (Section 5.2.5) and steam.

The use of steam (ie water vapour) is the best choice as this is an isothermal action that has the least disturbance on the temperature conditions. If misting is used there is a tendency to cool the air as the water adsorbs heat as it evaporates. Spray nozzles can in fact be used to dehumidify if the water temperature is below the dew point of the atmosphere.

Humidity in a growth chamber is most often controlled by a cooling coil to dehumidify together with electric heaters to maintain temperature. This system works well for temperatures between about 12-27°C but at high humidity, ice forms on the cooling coil if the ambient temperature is reduced to 10°C. This is because to dehumidify the air the cooling coil must be at or just below the dew point. For a temperature of 10°C this means that the cooling coil must be below freezing to dehumidify the air under 50% RH. It is this relationship between temperature and humidity that makes humidity control so difficult. Variations of 0.5°C in temperature give 3% fluctuations in the RH if there is no change in moisture content. To put it in perspective 80% RH at 30°C requires 250% more water in the atmosphere than 80% RH at 15°C.

∏ Use Figure 5.22 to decide what the vapour density is for a dry bulb temperature of 30°C and a wet bulb temperature of 20°C.

The reference points for these temperatures intersect on the 45% RH line which is equivalent to a vapour density value of about 12.5 g m^{-3} (read this from the left-hand scale).

5.5.5 Vapour pressure deficit

The difference between the saturated vapour pressure and the actual vapour pressure at any given temperature is known as the vapour pressure deficit (VPD). This value decreases as the relative humidity increases and is also temperature dependent (Table 5.18). As transpiration by plants is driven by the VPD. The recommended measure of humidity is VPD rather than relative humidity.

Temp (°C)	35	30	25	20	15	10
VPD(kPa)	0.55	0.42	0.33	0.23	0.17	0.12

Table 5.18 Changes in the vapour pressure deficit (VPD) in kPa with temperature, at 90% RH.

5.5.6 The importance of humidity to plant growth

Transpiration, the loss of water through the leaves to the external atmosphere, is vital to the plant for a number of reasons including cooling and providing nutrients etc via the flow of water induced through the plant. Therefore anything that restricts or interferes with transpiration is likely to have detrimental effects on the plant. Plants can transpire 100% of the water content of their leaves on a dry sunny day. Normally the water vapour generated would evaporate and be absorbed by the atmosphere. The situation in a greenhouse is somewhat different, however, as the humidity will have a tendency to increase. Let us examine some data relevant to this (Table 5.19).

| | Humidity treatment (VPD) | | | |
	0.1 kPa	0.2 kPa	0.4 kPa	0.8 kPa
leaf size (cm^2)	1159	1364	1817	2022
fruit yield (kg m^{-2})	10.1	10.7	11.1	11.4

Table 5.19 The effect of different humidity treatments (kPa of VPD) on leaf size (length x breadth cm^2) and yield (kg m^{-2}) of tomatoes. Adapted from Holder and Cockshull, J Hort Sci, 1990, 65.

∏ What is the most significant aspect of these findings? And what consequences does this have for the tomato grower?

There is a pronounced effect on the growth of leaves when the highest and lowest humidity treatments are compared. High humidities (low VPD) reduce leaf size by up to 42%, yet yield is only reduced by 11% at 0.1 kPa (compared to 0.8 kPa) and this falls to 6% at 0.2 kPa. This is mainly as a result of reduced fruit size rather than as a result of reduced number (data from other times of the year show almost no difference in yield). Consequently there is little point using costly humidity control trying to attain higher vapour pressure deficits than 0.3 kPa as the yield responses at lower humidities are minimal.

RH affects calcium

Another effect of high humidity is that the calcium content of the leaves is reduced. In tomato, a decrease in the vapour pressure deficit from 0.8 kPa to 0.1 kPa results in a 50% reduction in calcium content. Symptoms of calcium deficiency, chlorotic and bleached leaflet margins, appear at about 0.2 kPa in tomato but cucumber is more sensitive showing symptoms at 0.4 kPa.

5.5.7 Humidity and temperature control

Cooling can be effected by opening vents in the greenhouse when the internal temperature is too high (Section 5.3). This also has the effect of increasing the loss of water vapour from the greenhouse and therefore increasing transpiration.

∏ How will this effect leaf temperature?

We know that an increase in transpiration will have a cooling effect on the plant which consequently cools more rapidly than the greenhouse atmosphere. Figure 5.23 demonstrates that the cooling can be substantial, there is a drop of almost 10°C within a few minutes which can result in damage to the plant, particularly around the petiole. Sudden temperature changes can be avoided by opening the vents more slowly.

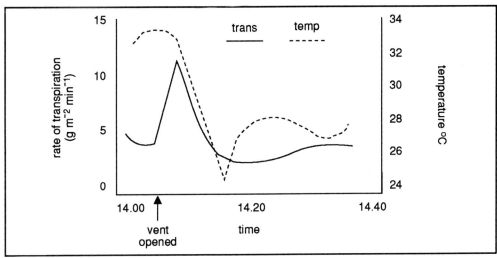

Figure 5.23 The effect on transpiration (g m^{-2} min^{-1}) and leaf temperature ($^{\circ}$C) of an increase in ventilator aperture from 0 to 60%. Adapted from Bakker, J C, 'Acta Hort', 1984, 148.

5.6 Diseases of greenhouse crops

The glasshouse is usually designed for the production of a small number of specialist crops, there is little scope for crop rotation and if the soil is not changed, pathogens can accumulate. Soil borne infections can be the most devastating and are perhaps the most difficult to control. Figure 5.24 shows the distribution of plant pathogens in the soil.

∏ What can you say about the distribution of pathogens in the soil from the information in Figure 5.24?

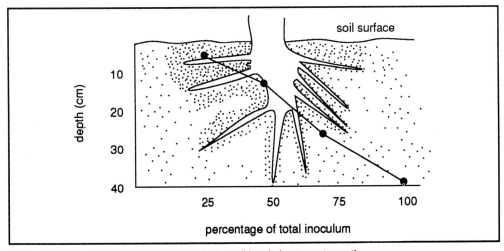

Figure 5.24 The distribution of pathogens in the soil in relation to root growth.

Perhaps not surprisingly, the distribution of pathogens follows the pattern of the roots and most of the organisms are found in the top 20 cm of soil. The spread of these pathogens can, in part, be controlled by the use of limited crop rotation, for example tomatoes and lettuce may be effective. The diseases of lettuce can be effectively controlled with fungicides, and tomatoes are grown on reduced treatments as their crop duration is less.

soil sterilisation   This strategy is unlikely to be sufficient in the long term and it restricts what the grower may produce. Treatment of the soil to remove pathogens and other harmful organisms is periodically required. The two most common methods are to heat the soil by steam or use a chemical such as methyl bromide.

Figure 5.25 shows the temperatures required to effectively sterilise the soil.

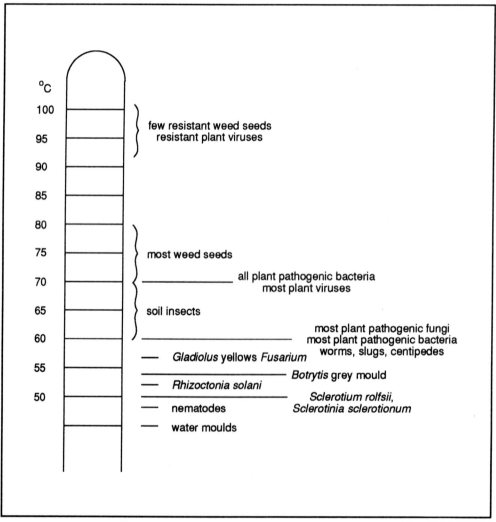

Figure 5.25 Temperature (°C) needed to kill plant pathogens and other harmful organisms. Adapted from 'Greenhouse Operation and Management', Nelson, P V, Reston Publishing Co, 1985.

Π   What temperature would you choose for the treatment of the soil in your greenhouse? See Figure 5.25.

You may well consider that 100°C is the temperature of choice as almost everything that is harmful to the plant will be destroyed at that temperature. In fact a temperature of about 75°C at a depth of 20 cm is routinely used.

Π   What advantage does this temperature have over the treatment at 100°C?

A balance must be struck between reducing the numbers of pathogenic organisms and retaining the beneficial ones like the nitrifying bacteria and other beneficial microbes. Also cost is important.

Steam can be introduced through pipes or spikes in the soil but the easiest method is to trap the steam beneath a PVC sheet. Steaming has the advantages over methyl bromide application of speed and lack of toxic residues; a period of one week is needed for chemical treatment and the soil must be checked for residues.

5.6.1 Cultivar resistance

TMV   The use of cultivars resistant to a particular pathogen(s) is another important method for the control of disease. The occurrence of TMV (tomato mosaic virus) in the UK is shown in Figure 5.26.

Π   From the information given in Figure 5.26, the introduction of resistant cultivars carrying the gene Tm-1 led to some reduction in the occurrence of TMV-strain 0. What was the effect of the introduction of these cultivars on the incidence of TMV-strain 1 infections?

The effect was that the incidence of infections by TMV-strain 1 increased. It would appear, therefore, that although gene Tm-1 bestowed resistance to TMV-strain 0, these cultivars were still susceptible to infection by TMV-strain 1.

Π   What gene appears to provide protection against TMV-strain 1 infections?

You should have identified gene Tm-2^2.

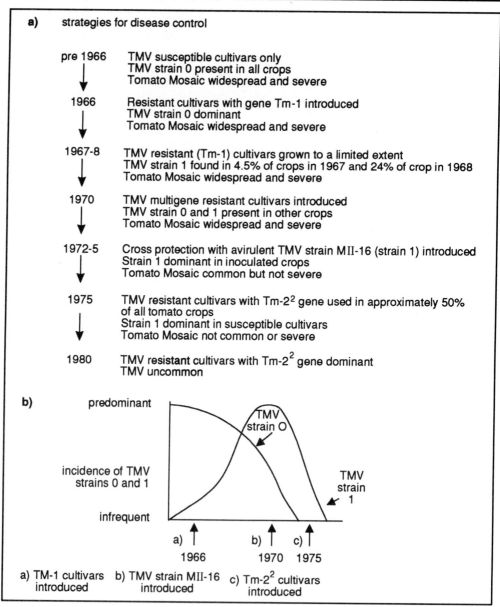

a) strategies for disease control

pre 1966   TMV susceptible cultivars only
           TMV strain 0 present in all crops
           Tomato Mosaic widespread and severe

1966   Resistant cultivars with gene Tm-1 introduced
           TMV strain 0 dominant
           Tomato Mosaic widespread and severe

1967-8   TMV resistant (Tm-1) cultivars grown to a limited extent
           TMV strain 1 found in 4.5% of crops in 1967 and 24% of crop in 1968
           Tomato Mosaic widespread and severe

1970   TMV multigene resistant cultivars introduced
           TMV strain 0 and 1 present in other crops
           Tomato Mosaic widespread and severe

1972-5   Cross protection with avirulent TMV strain MII-16 (strain 1) introduced
           Strain 1 dominant in inoculated crops
           Tomato Mosaic common but not severe

1975   TMV resistant cultivars with $Tm-2^2$ gene used in approximately 50%
           of all tomato crops
           Strain 1 dominant in susceptible cultivars
           Tomato Mosaic not common or severe

1980   TMV resistant cultivars with $Tm-2^2$ gene dominant
           TMV uncommon

b)   predominant

incidence of TMV strains 0 and 1

infrequent

a) 1966   b) 1970   c) 1975

a) TM-1 cultivars   b) TMV strain MII-16   c) $Tm-2^2$ cultivars
introduced   introduced   introduced

Figure 5.26 The incidence of TMV in the UK from 1966 to 1980. Tm-1 is a resistance cultivar of tomato introduced in 1967-1968 and Tm-2 was introduced in 1971.

We can interpret these observations in the following way.

The decrease in the incidence of TMV-strain 0 from 1967 to about 1970 as a result of the introduction of Tm-1 is very pronounced and demonstrates the effectiveness of resistant cultivars. The reason for the upsurge in the total TMV population is that TMV adapts quickly and a new strain of TMV (strain 1) to which Tm-1 was susceptible had been increasing steadily since 1966. This trend continued until a new resistant variety of tomato ($Tm-2^2$) was introduced.

SAQ 5.9   What do you think will be the likely occurrence of TMV from 1990 onwards?

Plants can also be inoculated with avirulent strains of virus that afford some degree of cross protection.

Grafting on to resistant rootstocks can also be successful but requires a considerable amount of labour and may allow less common diseases to develop because of incomplete resistance in the rootstock.

5.6.2 Avoidance

Reducing the possibility of diseases spreading by using 'clean' seeds and stock plants is essential to maximise production, especially as the most damaging effects occur early in plant development.

The use of nutrient film techniques (NFT) for the production of tomatoes and lettuce is a successful method for reducing diseases. There are few problems related to the use of NFT although *Oldipium brassicae*, a vector of lettuce big vein, is difficult to eliminate. Cucumber can be grown on straw bales, which has the added advantage of providing CO_2 and heat as it degrades. Barrier methods, such as the use of deep dug peat in bags or concrete troughs, will also reduce the incidence and spread of infection.

5.6.3 Environmental control

Plants are more susceptible to infection if they are kept under sub-optimal conditions and are subjected to stresses. Excessively humid conditions will encourage fungal infections particulary when moisture forms on the leaf surface, so it is particularly important to maintain the leaf surface above the dew point which will keep leaf infections to a minimum. Temperature control is not usually an option as the optimum for crop production coincides with the infection optimum, however a brief increase in temperature that has little effect on the plant can sometimes be used. A temperature of 25°C for 3 days dramatically reduces the incidence of *Verticillium albo-atrum* (V Wilt) for example.

It should also be remembered that resistance can depend on the temperature. Tomato variety Tm-1 is resistant to 53% of isolates of TMV at 17°C but only 34% at 30°C. Improved control of the environment will have beneficial effects by reducing the opportunities for infection and spread of disease as well as improving productivity. This degree of control is becoming possible through the use of computerised systems, however more information is needed about the biology of the plants and diseases before the full potential can be realised.

5.6.4 Economics of control

A knowledge of the effect of infection on the final yield of a crop is necessary to determine whether or not control measures will be economically viable. Leaf mould of tomatoes, caused by *Fulvia fulva*, has little effect on yield unless at least 50% of the plants' leaf area has been covered for about 6 weeks. If the infection is less intense than this, the cost of treatment may be more than the added value which would be derived from the crop. In general, low levels of infection may be tolerated until the loss in revenue justifies the use of control measures.

The continued presence of infectious agents within the greenhouse may cause problems with subsequent crops as there is a greater possibility of carry over of infections from one crop to another. Plants are not the only consideration either, the spores and other factors released by the microbes may cause allergenic reactions in the work force. All these factors must be balanced before deciding on the most appropriate course of action.

Summary and objectives

In this chapter we have examined the use of greenhouses and growth rooms for cultivating plants. We began by considering the major environmental factors which influence plant growth and the methods and devices we can employ to control these factors. In this context we considered temperature, light, carbon dioxide and water. In each case we explained how the environment in which plants grow within a greenhouse may be substantially different from that encountered in fields. We also showed how adjustments to one parameter may dramatically alter other parameters. Throughout, we have used experimental data to illustrate the issues described in the text. We completed the chapter by briefly considering the occurrence and control of plant diseases in greenhouses. We included a brief discussion of the physical and chemical methods that may be used to control disease outbreaks and pointed out the importance of producing disease-resistant cultivars.

Now that you have completed this chapter you should be able to:

- explain the importance of monitoring and controlling the temperature of a greenhouse;

- describe the strategies for ensuring even heat distribution;

- calculate the required heat input into a greenhouse from supplied data;

- explain why the methods used to measure light produce data that is directly related to the value of the light active in photosynthesis;

- explain why care has to be shown when selecting lamps to be used as an artificial light source and why often a mixture of lamps is used;

- explain how the rate of photosynthesis is influenced by the concentration of CO_2 and how enrichment of the atmosphere of a greenhouse with CO_2 may have beneficial effects;

- list some examples of air borne pollutants which may inflict injury on plants within a greenhouse;

- explain what is meant by absolute and relative humidity and describe the methods available for measuring relative humidity;

- explain what is meant by vapour pressure deficit (VPD) and describe the effects of increased VPD on crop growth and yields;

- describe the reasons why plants in greenhouses may be vulnerable to attack by pathogens and pests and explain the precautions and strategies that may be employed to reduce disease outbreak.

Hydroponic vegetable production

Hydroponic vegetable production

6.1 Introduction

Earlier chapters in this text have described the production of the world's food through what might be called the standard agricultural procedure of growing crops in soil in fields. Mankind has been interested in modifying the environment for the growing of crops since the time of the Romans in the first century AD, and, with our increased understanding of the control of plant physiology and development, we now have the knowledge to practice controlled-environment agriculture. The term, controlled-environment agriculture (CEA) encompasses all aspects of modifying the natural environment to optimise plant growth. It covers the whole spectrum of operation from that of a simple greenhouse, in which there is control only of the minimum temperature, to that of a true controlled-environment facility in which light, temperature and humidity are provided artificially and in which CO_2 enrichment can be easily achieved. In CEA, many crops are grown using one form or another of hydroponics, a system for growing plants in nutrient solutions with or without the use of an artificial medium, for example sand, perlite or rockwool.

controlled-environment agriculture (CEA)

In this chapter we will describe the various ways in which hydroponics is used for crop production and compare the performance of hydroponically grown crops against soil-grown crops. Although hydroponics can be used to grow flowers, foliage and bedding plants, we will concentrate on the production of vegetables. Although most hydroponic systems are used in greenhouses, this is not essential and these systems can be used in the open.

6.2 Hydroponics can use liquid or aggregate systems and be open or closed

In liquid hydroponic systems the roots of the plant are bathed directly in the nutrient solution, and there is no supporting medium for them. Aggregate systems utilise a particulate material which provides support for the roots. The aggregate is usually, but not always, inert. A separate classification is concerned with the manner of use of the nutrient solution. In an open system, once the nutrient is delivered to the plant roots it is not re-used. In a closed system the nutrient is recovered and recycled.

6.3 Hydroponics does have some disadvantages

As with any innovation of growing systems there are disadvantages as well as advantages. The advantages include the fact that crops can be grown where no suitable soil exists. The control over nutrient supply and the partial control over temperatures and, perhaps, CO_2 availability, allows plants to be grown at higher densities than in open-field agriculture (OFA). A major advantage is the fact that the plants are grown isolated from the underlying soil, which avoids the need for expensive sterilisation

hydroponics in comparison with OFA

programmes. This also allows a quicker turnaround from one crop to the next. The use of hydroponics in an enclosure also allows better control of diseases.

There are two principal disadvantages to hydroponics. One is the high costs of capital and energy inputs compared with OFA, and the degree of competence in plant science and in engineering skills required for successful operations. Because of the costs, hydroponics is limited to the production of crops with high economic value, and to production in specific regions, and to production at times of the year when OFA crops are not available. Hydroponics is not suitable for agronomic crops, eg cereals. In 1977 it was considered that market prices for agronomic crops would have to increase five fold for hydroponics production to break even. Since then, the costs of capital and energy have increased by much more than the market price of agricultural commodities. Pricing studies reveal that only high-quality, garden-type vegetables eg tomatoes, cucumber and speciality lettuces, can provide suitable returns on investment. These are the major crops of hydroponics today, with eggplants, peppers, strawberries and herbs fulfilling minor roles.

high costs of hydroponics limits use to particular circumstances

The utilisation of hydroponics for the production of vegetables shows some interesting variations on a world basis. It was estimated that there was approximately 300 ha of hydroponic cultivation systems in Europe in 1990, but less than 300 in the USA. The reasons for this are likely to be complex but include the fact that the diverse climate of the USA allows conventional field production of vegetables somewhere on the continent during any time of the year, which, coupled to the existence of a rapid and effective transportation system, makes if difficult for hydroponic vegetable production to compete with field crops.

hydroponics in Europe and the USA

6.4 Preparation of plants for transplanting into hydroponic systems

All of the systems to be described in this chapter involve the transplanting of young plants from their original growth medium into the apparatus to which the hydroponic solution is applied. These young plants are grown in ways which differ little from standard horticultural practice. This involves sowing seeds at relatively high density and transferring the germinated seedlings to a lower density growing system when the cotyledons or first leaves are fully open. In Britain, this process is referred to as 'pricking out'. The pricked out seedlings are then allowed to grow, usually for 3-4 weeks, before transplanting into the hydroponic system. During the pre-hydroponic stage plants are often grown in peat/sand mixtures and, depending on the type of hydroponics, excess growing medium is usually washed off at transplanting to minimise the transfer of solid into the nutrient solution. Young plants can be obtained with their roots free of adhering particles by growing them in smooth course gravel.

sowing and pricking out

6.5 Liquid hydroponic systems come in a number of variations

Liquid hydroponic systems are, by their nature, closed systems in which the plant roots are exposed to the nutrient solution in the absence of an aggregate root-supporting medium, and the nutrients are returned to a reservoir and re-used. The system could, technically, be used in open mode, but it would be very wasteful of nutrients. The nutrient film technique (NFT) is the basic system but several modifications of this have been employed.

liquid hydroponic systems are usually closed systems

nutrient film technique

6.5.1 The nutrient film technique

This was first devised by Dr Allan Cooper at the (then) Glasshouse Crops Research Institute (now Horticulture Research International), in the late 1960s. In NFT, a thin film of nutrient flows through channels which contain the plant roots. The roots spread out laterally in the film, forming a dense but shallow root mass (Figure 6.1). This results in very good root aeration.

Nutrient solution is pumped to the higher end of each trough and flows by gravity to the lower end, where it is collected and returned to the nutrient reservoir. The trough is lined with 600-800 gauge white on black plastic which is drawn together around each plant to exclude light and reduce evaporation. This gauge of plastic gives a smooth base for even spreading of solution. The plastic is arranged so that the black is on the inside and the white outside, the raised sides being stapled or pegged forming a triangular or circular cross section (Figure 6.1 a).

SAQ 6.1 Can you think of a reason why white on black plastic is used, as described above, rather than, for example, clear plastic?

A capillary mat (Figure 6.1c) may be used to line the plastic channel and this ensures good contact between the nutrient solution and the roots of the initial transplant.

The slope of an NFT channel needs to be between 1 in 50 and 1 in 75 and the channels can be laid directly onto a carefully graded floor of the greenhouse (Figure 6.1 b) or may be raised and supported by rigid troughs (Figure 6.1 d). If the channels are to be laid directly onto the ground, this must be carefully sloped to provide the required gradient and must obviously avoid depressions which would cause the formation of stationary pools of nutrient.

SAQ 6.2 You should be able to give two reasons why stationary pools of nutrient may be detrimental to plant growth in an NFT system.

operating temperatures are important and can influence yield

The use of raised troughs removes the need to slope the floor of the greenhouse; a level greenhouse floor is slightly easier to work in than a sloping one. Typically plants are grown at a density of approximately 30 m^{-2}. The volume of solution circulating around NFT systems is relatively small and this permits the easy modification of its temperature independently from that of the air compartment of the facility. After the energy crunch of the early 1970s, operators experimented with the use of reduced night-time air temperatures, as an energy saving measure. These experiments led to a reduced yield of early tomatoes but an increased overall yield. Early tomatoes, of course, carry a considerable price premium, but it was found that the reduction in early yield was partially counteracted by heating the NFT solution. Winter lettuce appear to be affected similarly. In desert regions the cooling of the nutrient solution has been shown to have beneficial effects. 35°C appears to be a working maximum temperature for plants in hydroponics generally, although temperatures of 25°C and above stimulate bolting in lettuce. There is also evidence that cooling the nutrient solution reduces the incidence of damping off caused by the fungus *Pythium aphanidermatum*. Whereas much is known about the effects of temperature on plant physiology, in general, insufficient information is to hand to allow a full explanation of the effects just described.

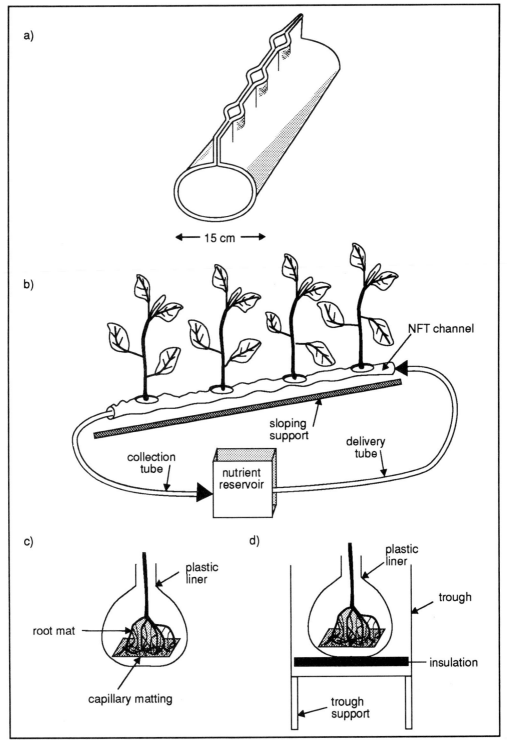

a)

← 15 cm →

b)

NFT channel

sloping support

delivery tube

collection tube

nutrient reservoir

c)

plastic liner

root mat

capillary matting

d)

plastic liner

trough

insulation

trough support

Figure 6.1 Main features of a NFT system. a) Flexible black/white plastic channel, folded and stapled to form a tube with the white side outside, b) general layout of a system, c) cross section of unsupported channel, d) cross section of supported channel. Note we have over-emphasised the slope - it is normally only about 1 in 50 to 1 in 75.

6.5.2 Variations on the NFT theme

Covering a greenhouse floor with cast concrete shaped into NFT channels is one example of a modified system. The surface of the concrete is sealed with epoxy resin and the nutrient solution applied directly to the channels. This system has been used mainly for the production of lettuces and here a rigid plastic support is used, with holes for the plants to be threaded through (Figure 6.2). An extension of this idea is the use of movable channels supported on low sloping benches. The channels are separate from each other and their distance apart can be changed. The channels are placed close together when growth is initiated and can be moved apart to give more growing space as the plants grow.

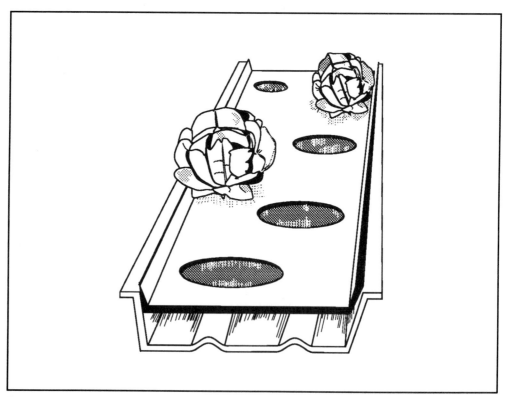

Figure 6.2 Lettuce can be supported on rigid plastic sheets suspended above the channel through which the nutrient flows.

mechanical
harvesting
device

This system lends itself readily to mechanical harvesting. In a system developed in the Netherlands lettuces, are planted in flexible plastic holders. At harvest time a winching machine pulls the plastic holders, lettuce included, up an incline and onto a spool. Just before the spool, the lettuce stems are mechanically cut close to the plastic and the heads diverted onto a conveyor belt, while the roots are brushed off onto another (Figure 6.3). Such a system obviously depends on the production of uniform crops.

Two systems using pipes have been devised. In one, seedlings are planted in holes in 3 cm diameter pipes held at a slope of 1 in 30 (almost twice that used for conventional systems). This arrangement significantly increases the usable growing surface and accommodates 40 plants m^{-2}.

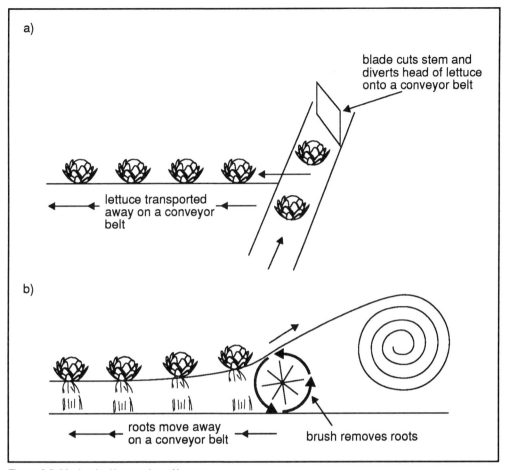

Figure 6.3 Mechanical harvesting of lettuce.

A vertical pipe system has also been developed. Here, plastic or aluminium pipes, 1.3 m long and 0.15 m wide, are suspended vertically in the growing area. When used for lettuces, 20-28 plants are inserted into holes in the tubes. Nutrient solution is pumped to the top of each pipe and allowed to trickle down over the plant roots. This system has been used in Israel for the production of vegetables and flowers, but is not widely used elsewhere.

6.5.3 Floating hydroponics

There have been numerous reports of the use of floating hydroponic systems over the last twenty years or so, dealing mainly with the production of seedlings for transplanting. A small number of studies have been carried out on the production of lettuces by the system and we will examine one example, that conducted by the University of Arizona in 1981-1982. The basic system consisted of nutrient channels, referred to as raceways, 4 m wide by 70 m long by 30 cm deep, on which were floated large sheets of 2.5 cm thick expanded polystyrene. Lettuce seedlings were transplanted through holes in the polystyrene 15-20 cm apart to give approximately 33 plants m^{-2}. There were two particular advantages to the system. The nutrient pools were virtually frictionless conveyor belts for planting and harvesting movable floats. As a crop of floats was harvested at one end, new floats with transplants were introduced at the other, the floats being very easy to move by hand. This led to the second advantage

use of expanded polystyrene trays

which was that only narrow walkways were required between the raceways, which increased the proportion of the greenhouse actually being used for growth. The greenhouse used in this study was 0.5 ha in area and produced lettuces at a rate of 4.5 million ha^{-1} year^{-1}, but there are no signs that this system is being used commercially.

6.5.4 Hydroponics can also be used as a mist

root mist techniques

A number of systems have been described in which the nutrient solution is applied as a mist, either to the roots alone or to the whole plant.

In the root mist technique, plants are grown in holes in horizontal polystryrene panels with their roots suspended in air inside a spraying box, into which the nutrient is applied as a mist (Figure 6.4). In early systems the nutrient mist was applied continuously, but studies have shown that this is not necessary. In one study, cucumbers were grown from the seedling to the fruiting stage in 40 days on a regime of 7s of mist in each 10 mins. Under such conditions the roots were kept moist but also well aerated.

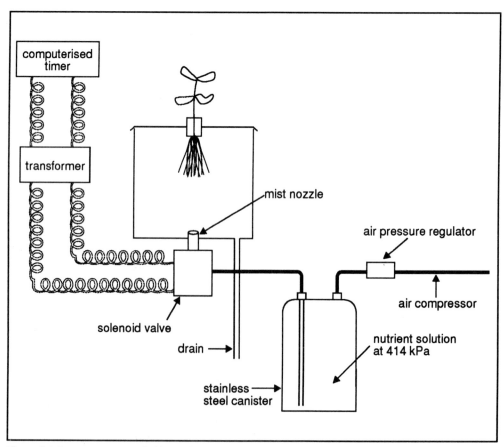

Figure 6.4 A computer-controlled intermittent mist applicator. (Re-drawn from Peterson LA, and Krueger AR, Crop Science, Vol 28 p712-713).

aeroponics

Systems using this method of delivery, sometimes called aeroponics, have been developed in Italy and the USA. The plants are grown in polystyrene sheets as in floating hydroponics but, because the nutrient is sprayed out, these need not be horizontal. Indeed the commercial systems utilise what is called an A frame (Figure 6.5) in which the polystyrene sheets form two sides of an end view equilateral triangle.

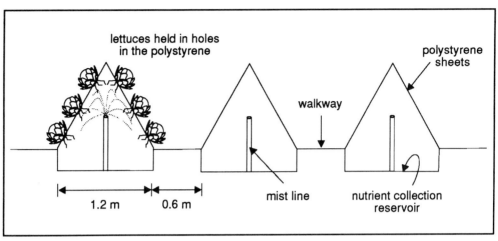

Figure 6.5 End view of three A frames showing the inclined polystyrene panels.

The most uniform growth is obtained when the inclined panels are oriented so that they face east-west.

∏ Assuming that the number of lettuce per unit area in the panels shown in Figure 6.5 is the same as that described above for the floating hydroponic system, and that growth rate is the same, what is the approximate potential increase in yield per unit area of greenhouse in the A frame aeroponic system compared with that using floating hydroponics?

The approximate potential increase in yield will be 33%. The A frame results in a doubling of the yield over the 1.2m of equivalent horizontal space but the non-productive inter-frame space is equal to half of the productive area.

$$\text{Thus yield} = \frac{2}{1.5} = 1.33$$

SAQ 6.3   Is it feasible to do away with the inter-frame walkway in Figure 6.5 so as to obtain the maximum doubling of yield?

potential of using mist cultures in space

Another potential commercial application of aeroponics is the production of vegetables in locations with restrictions in area and, or weight, such as large, manned space craft designed for long occupancy. Merle Jensen, of the Experimental Research Laboratory of the University of Arizona has designed an aeroponic system in the form of a revolving 'space drum' which simulates extra terrestrial gravity and these ideas are being developed by the Boeing Company.

commercial application of mist cultures

The extension of the misting system to its logical extreme, that of applying nutrient mist to the whole plant, has been done by two companies in Britain. Although only a very limited number of species of plant has been grown successfully in this way, the method is intriguing and appears to have high potential. The two companies involved have developed a method for the production of what is called barley grass. Barley seed is germinated, grown for seven days and then harvested for cattle feed. The plants are grown in custom-made chambers, the largest of which resemble caravans, in which temperature is controlled and where light is provided artificially. Trays are half filled with barley seed which is placed in the misting unit. The seed absorbs nutrients from the mist (applied for 30 seconds every six hours) germinates and, over the seven day period, grows to a height of approximately 30 cms (Figure 6.6).

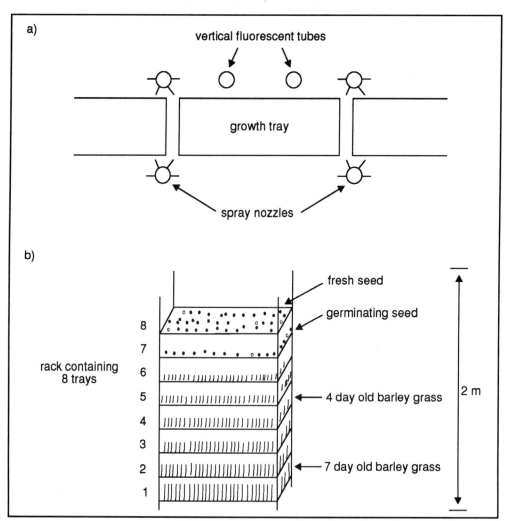

Figure 6.6 Design of unit for production of barley grass. a) Plan view of trays. Light is provided by fluorescent tubes in the wall and in the ceiling. b) Side view. Each day the tray is removed from position 1 and the grass harvested, trays in positions 2-8 are moved down 1 position and a tray with fresh seed placed in position 8.

barley grass only commercially viable in a limited set of circumstances

The roots of the barley intertwine with each other and produce a dense mat. At harvest the whole of the contents of the tray, roots and all, are removed and fed to livestock. The largest units, approximately 45 m³, produce 1 tonne of barley grass per day at a cost in 1988 judged to be of the order of £23 per tonne. The economics are such that the units are used only in situations where farm land is not available or very expensive, as in Japan, or where the animals to be fed warrant the special treatment. The British horses which competed in the 1988 Seoul Olympic Games were fed predominantly on barley grass produced in the manner described above. Although company literature refers to the production of other crops using this system, no information about productivity is available.

6.6 Aggregate Hydroponic Systems

In aggregate systems a solid, usually inert, material provides support for the plants. Whereas most liquid hydroponic systems are closed, most aggregate systems are open ie nutrients are not recycled. The spent nutrient should be collected and disposed of in an environmentally acceptable way and, in most cases, this involves utilising it to irrigate field grown crops.

6.6.1 Numerous growing materials are available

aggregates used include sand, gravel, peat, rockwool perlite and gels

Because the nutrient is not recycled, open systems are less sensitive to the composition of the medium, particularly the quality of the water used. This has led to experimentation with a wide variety of aggregates which include sand, gravel, peat, rockwool and perlite as well as some recently developed synthetic materials, such as the polyacrylamide 'Growgel' and a foam made from urea formaldehyde. The choice is partly governed by location and by the local availability of materials.

sand and gravel can be re-used indifinitely

Sand is readily available in desert regions and the University of Arizona has conducted numerous experiments using pure sand and various mixes containing rice hulls, pine bark and peat. The main advantage of sand, and gravel shares this, is that it does not undergo physical breakdown and, therefore, does not require to be replaced every 1 or 2 years. Mixes of sand with degradable materials would not share this advantage.

Peat has been used for many years, generally in conjunction with non-degradable aggregates such as sand which improve its aeration, but its use in hydroponics has been almost totally superseded by alternatives.

rockwool needs to be pre-treated

Rockwool was first developed as an acoustical and insulation material and is made from a mixture of diabase limestone and coke by melting at temperatures in excess of 1500°C, feeding onto a spinning drum which spins it out into threads, and pressing into sheets. This material is not, itself, suitable for horticultural use but must be treated in a way which has remained confidential.

Perlite is mined as a mineral which, when treated at high temperature, vaporises water inside it to form air pockets in a manner analogous to the production of expanded polystyrene.

The various synthetic materials are all offshoots of the insulation industry. As each new material is developed, it is tested for possible horticultural use, with or without further modification. Although trials are continuing with these types of material they are not yet widely used in production systems.

Aggregate materials are used in a variety of ways which we will now examine.

6.6.2 Trough or trench culture

arrangement of troughs or trenches depend upon the crop

Aggregates may be used in trenches made into the ground or in troughs laid on the surface; in both cases plastic sheets are used to retain the nutrients. The sheets are normally laid as a double layer to prevent leakage. The size and shape of the beds shows wide variations; tomatoes are usually grown in beds wide enough for two rows to aid management and harvest. Lettuce and other low-growing crops are grown in wider beds, limited usually by the reach of the average operator.

In desert areas, however, it is common to cover the whole of the greenhouse floor with plastic and then backfill the area with sand. All variations on this theme require that there should be a slope of approximately 15 cm per 35 m for good drainage and **drip irrigation** well-perforated drainpipes should be laid at the base of the aggregate. Nutrient is provided by a tube applied to each plant, referred to as drip irrigation.

6.6.3 Bag culture

This is similar to trough and trench culture except that the aggregate is contained within bags, thus avoiding the cost of trenches etc. The bags are used for 2 years and then re-sterilised, depending on the aggregate; it is easier to sterilise the bag's contents than bare sand. Bags are usually black on the inside and white outside. Bags, usually of 50-70 litre capacity, are laid flat on the ground and holes made in the top surface for the plants. Slits are made low down on the sides for drainage and nutrients are supplied by drip irrigation. An alternative system has the bags sitting on plastic lined gullies set on a gentle slope, with 'weirs' made of polystyrene rods set up to create a series of reservoirs. The bags are placed in the reservoirs and slit vertically on the lower edges to allow entry of the nutrient (Figure 6.7).

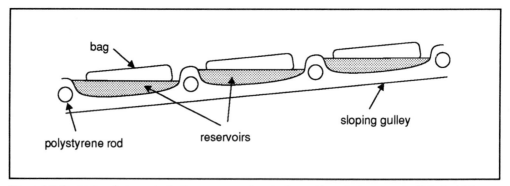

Figure 6.7 Formation of reservoirs by the presence of weirs formed by polystyrene rods. Note that this figure is somewhat stylised. The sloping gulley is usually quite shallow (1 in 50 to 1 in 75).

Bag culture utilises mainly perlite and the system was developed by the Soil Science Unit of West of Scotland Agricultural College. It remains the method of choice in Scotland and Northern England and in Eastern Europe, for example Bulgaria.

6.6.4 Slab culture

This uses predominantly rockwool in slabs which are usually 15-30 cm wide, 75-100 cm long and 75 mm thick. Figure 6.8 shows two ways of preparing seedlings for growth in rockwool slabs.

SAQ 6.4

In the production of tomatoes, method b) in Figure 6.8 has been found to lead to earlier establishment of the transplant in the rockwool slab which leads to earlier production of the first flowers. Can you suggest an explanation for this?

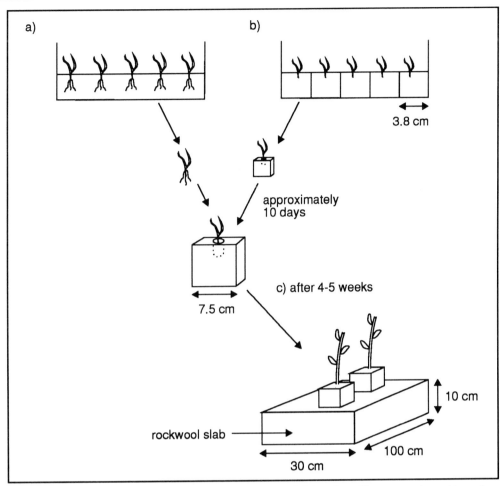

Figure 6.8 Growth of seedlings and transplanting into rockwool slabs; a) seedlings grown, in peat based mixture for approximately 10 days, are pricked out into a cavity in a 7.5 cm x 7.5 cm x 7.5 cm rockwool cube and bedded in with rockwool granules; or b) seeds are germinated individually in 3.8 cm x 3.8 cm x 3.8 cm rockwool cubes and transferred into the cavity of the larger cube. c) 4-5 weeks later the larger cube is placed on top of the rockwool slab. Prior to placing on the slabs, plants are kept watered by the application of the nutrient solution.

Typically the greenhouse floor is covered with white plastic and the slabs, usually arranged in rows of two, are individually wrapped, but not sealed, with white plastic. The slabs are arranged to have a slight inward tilt towards a central drainage zone and may be supplied with bottom heat through hot water pipes (Figure 6.9).

rockwool is now the support of choice for seedlings

Since the advent of rockwool, it has become the chosen support for seedlings prior to their incorporation into a wide range of hydroponic systems and also become very popular as an aggregate. Rockwool has a massive pore space which can hold both water and nutrients for plant growth. Typically the fibre occupies only 5% of the volume of a slab and its structure is such that it is impossible to overwater it. A standard rockwool slab with a volume of approximately 7 litres will contain approximately 1 litre of water or nutrients even when fully drained. It is also very easy for plants to extract water from it. Plants can remove up to 90% of the water in a slab before there is any significant increase in resistance, which is much better than most other substrates such as soil or peat. Because free-draining slabs will never become waterlogged, it is very easy to maintain slabs at the optimum water/air ratio which is between 65/35 and 50/50 water/air. The mass of roots of any crop is likely to be greater in a well managed rockwool slab than in almost any other medium because the conditions are so consistent. This means that water and nutrient uptake are more likely to be close to the optimum levels all the time.

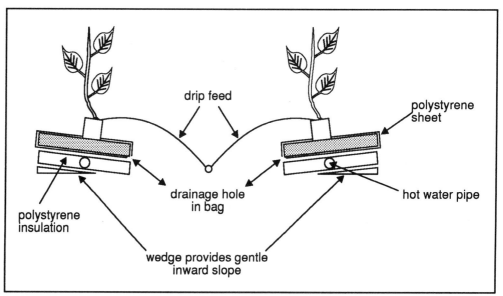

Figure 6.9 Layout of slabs for rockwool culture with bottom heat. Before seedlings are transplanted the slabs are soaked with nutrient solution. Plants in the rockwool cubes are simply placed on top of the slabs through holes cut in the plastic. In a well managed system roots will grow down into the slab within 2-3 days. Note that again we have rather exaggerated the slope of the slabs. The slope usually is as low as 1 in 30 to 1 in 50.

6.6.5 Closed aggregate systems are sometimes used

Closed systems using gravel as the support material were commonly used during the early days of hydroponics but these have been superseded by the open methods, due to the advantage of the latter in avoiding the build up of toxic compounds. The typical system did not circulate the nutrient, as in liquid NFT. Rather, the nutrient was pumped into the gravel bed to within 2 cm of its surface, and then allowed to drain out by gravity. Such systems had high capital outlays because of the need for leakproof beds and below-ground-level mechanical systems.

A more recent development is a combination of NFT and rockwool culture, which qualifies as a closed aggregate system. Here plants are established in rockwool slabs positioned in channels containing the cycling nutrient. This system uses less rockwool than the open system described above.

6.7 Choice of nutrient solution

Many formulae have been devised for hydroponic nutrient solutions and the only thing they have in common is that they all supply all of the essential elements: N, O, P, K, Cl, Mg, Ca, Fe, S, Mn, Cu, Zn, Mo and B. Some of the formulations are shown in Table 6.1

	Hoagland and Arnon (1950)	Resh (1981)	Verwer (1976)
N	210	175	173
P	31	65	39
K	234	400	280
Ca	160	197	170
Mg	48	44	25
S	64	197	103
Fe	0.6	2	1.7
Mn	0.5	0.5	1.1
Cu	0.02	0.03	0.017
Zn	0.05	0.05	0.25
B	0.5	0.5	0.35
Mo	0.01	0.02	0.058

Table 6.1 Composition (mg l^{-1}) of hydroponic nutrient solutions.
Hoagland, D.R. and Arnon, D.I. (1950) The Water Culture Method For Growing Plants Without Soil, Cir. 347 Calif. Agricultural Exp. Station, University of California.
Resh, H.M. (1981) Hydroponic Food Production, Woodbridge Press Publishing Co, Santa Barbara, California.
Verwer, F.L. (1976) Growing Horticultural Crops In Rockwool And Nutrient Film, p107-119. In: Proc. International Working Group or Soilless Culture, Las Palmas, Canary Isles, Spain.

SAQ 6.5

One essential element is missing from Table 6.1. What is it and where do you think the plant gets it from?

Π Examine the amounts of elements shown in Table 6.1. What is the major difference between Resh's medium and the other two?

Resh's medium contains approximately 40% higher mineral content.

Some operations use two concentrated nutrient stock solutions, one containing calcium sulphate the other contains the other minerals. The two separate solutions are necessary to prevent precipitation of the mineral concentrates. The maintenance of two separate stock solutions is not necessary if lower salt concentrations are used.

No one formula is inherently better than any other, as long as extremes of any single element is avoided, especially if the rate of supply is governed at least partly by the rate of consumption. This of course, is much more important in closed liquid systems. It is checked in two ways. Measuring electrical conductivity (EC) gives an indication of overall mineral content which is important because of its relation to plant water status. Periodically the content of individual elements should also be determined.

electrical conductivity indicates overall mineral content

recommended monitoring responses

Recommended assessment times are daily for pH and EC, at 2-3 weeks for macro-elements and 4-6 weeks for micro-elements. Many operators, particularly of small closed systems, not surprisingly find such measurements difficult to accommodate. A common practice is to begin with fresh nutrient solution, then at the end of the first week add one half of the original total nutrients, and at the end of the second week get rid of all the nutrient solution and start again with a fresh batch.

In open systems the nutrient solution is not recycled so it does not need to be monitored unless automatic methods are being used to add minerals to the irrigation water. It is necessary, however, to monitor the growing medium regularly because the rate of water use is much higher than the rate of nutrient use and mineral concentrations increase considerably. Tests show that not only the position of sampling but also the timing of it must be taken into consideration (Table 6.2).

	Conductivity (μS)	
	between plants	below plants
before application of drip irrigation	3400	2830
after application of drip irrigation	2500	2100

Table 6.2 Conductivity of rockwool slabs provided with a periodic drip feed of nutrient with a conductivity of 1800 μS.

SAQ 6.6

What recommendations would you make for a sampling protocol for an open system using rockwool slabs?

The UK Ministry of Agriculture Fisheries and Food (MAFF) recommendations (ADAS, P 3174, 1988) for sampling are that 15-20 slabs are chosen at random and that half are sampled between plants and half beneath plants and sampling should be at the single chosen time which is just before or just after irrigation. The slabs should be marked so that sampling occurs at the same slabs each time.

SAQ 6.7

In Table 6.2 rockwool slabs had been irrigated with solution with a conductivity of 1800 μS. Why did not the slab conductivity fall to this value and what could you do to bring it down to this value?

The ability to flush out chemicals is one of the advantages of the open rockwool system, but it must be remembered that the only way to reduce the conductivity of the solution is to reduce the nutrients. The water used will have sodium and chloride in it and these often contribute to the increase in conductivity of slabs in open systems. Whereas crops are able to withstand exposure to significant levels of sodium and chloride it is something to bear in mind. For this reason operations often maintain a supply of clean rainwater.

use of
rainwater

6.8 Effects of conductivity on productivity

An individual crop does not necessarily need a uniform conductivity throughout its life and different crops often require different conductivities.

The recommended protocol for tomatoes (MAFF, ADAS, P3174, 1988) is shown in Figure 6.10.

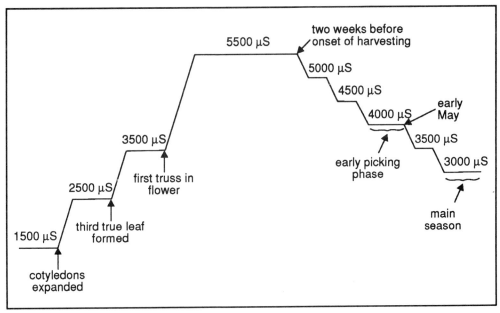

Figure 6.10 Recommended variation of conductivity during growth of tomatoes in open rockwool systems. (Data from MAFF, ADAS P3174, 1988).

| **SAQ 6.8** | It is considered that the level of nutrients is adequate for growth during the whole of the period indicated in Figure 6.10. What is the effect on growth of the increase in conductivity during the preflowering period? |

reduced vegetative growth improves flower setting

The increased conductivity during the preflowering period does, indeed, reduce plant growth. The reduced vegetative growth has been found to improve the setting of the first truss leading to an increase in early yield The gradual reduction in conductivity prior to the onset of picking is needed to obtain normal sized tomatoes. The conductivity is usually kept at this level for the remainder of the season, although it is sometimes lowered during the period of high transpiration rate in mid summer.

The tomato is classed as a vine and will grow to considerable heights if supported. This means that in glasshouses the plant soon reaches the roof and can suffer scorching if not attended to. A variety of systems are employed to avoid this; these are shown in Figure 6.11.

up and down, Dutch hook and layering systems

In the up and down method the plant is trained over a wire 2.10 m above the ground, allowed to grow down until almost reaching the ground and then trained up again. In the Dutch hook system the plant is trained along the wire. The layering system allows the plant to reach the horizontal wire, at which time the vertical line is moved 30-60 cm along the horizontal wire. This causes the base of the stem to bend over and lowers the shoot apex by 30 cm or so. When the stem grows and reaches the wire again, the process is repeated. Interplanting involves excising the apex of the early grown plants to stop growth and placing younger ones between them. All systems are labour intensive, made worse by the fact that the axillary shoots have to be regularly cut off. Older leaves are also removed. The layering system intercepts light most effectively of all the systems, probably because it maintains an approximately 2 m portion of actively fruiting stem in the vertical position.

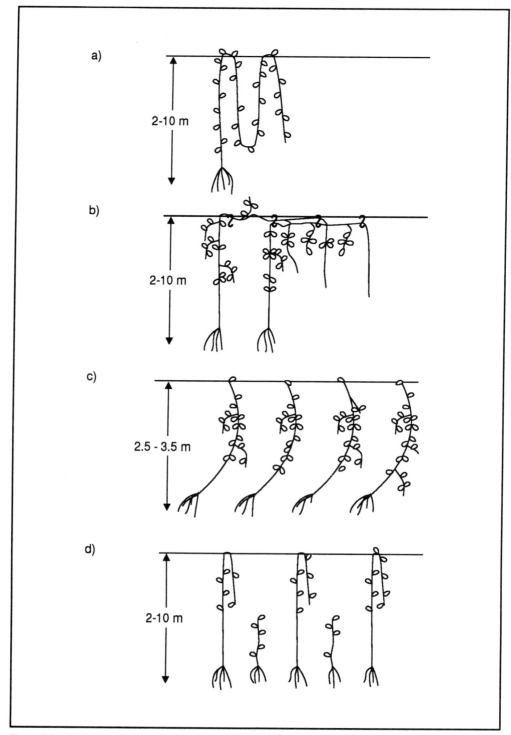

Figure 6.11 Training systems for tomatoes. a) Up and down, b) Dutch hook, c) layering and
d) interplanting. In each system, the plants are supported by chord or wire lines. (Redrawn from van de
Vooren *et al* (1986), Glasshouse Crop Production. In: The Tomato Crop, JG Atherton and J Rudich eds,
Chapman and Hall, London, 581-624).

Lettuce has been grown successfully on solutions with conductivities of 1500-1800 μS but their leaves become hard and leathery at higher conductivity levels and the proportion of marketable yield falls.

Cucumber grows less well than tomatoes in NFT. It suffers a lot of root dieback, thought to be caused by the poor nutrient distribution in the dense root mats typical of the vegetable. It performs well in rockwool and appears able to successfully utilise a wide range of conductivities.

6.9 Use of mycorrhizae

few studies on mycorrhizae in hydroponic systems

The roots of many species are infected with soil fungi forming the association known as a mycorrhiza. (see Biotol text, 'Crop Physiology', Chapter 4). It is, perhaps, a little surprising that only a small amount of work has been done to investigate the possible benefits of inoculating hydroponics systems with spores of mycorrhizae. However, some work on this topic has been done and, although no major breakthroughs of productivity have been noted so far, an interesting spin off has been achieved. The roots of hydroponically grown plants can easily be infected with spores of mycorrhizae and the fungus spreads very quickly through the system. If a closed liquid aggregate system is used, with expanded clay particles as the aggregate, the clay particles are found to contain high levels of spores and fungal mycelium and provide an excellent source of inoculum of the fungus. The particles can easily be decontaminated from infestation with various pathogens eg *Pythium spp* without serious loss of mycorrhizal infectivity and at least one company has brought such a product to the market place.

6.10 Nutritional disorders can occur in hydroponics

common problems are too much NH_4^+ and zinc and too little Ca^{2+} and K^+

None of the nutritional disorders noted in hydroponics are specific to it and they are more likely to occur in closed than open systems due to unwanted elements in the recycled liquid or in the chemicals used to prepare the nutrient solution. The most common problems are due to too much ammonium nitrogen and zinc and too little calcium and potassium. Less than 100 mg kg^{-1} of potassium in the nutrient solution reduces the percentage of high quality tomatoes and calcium deficiency can lead to blossom end rot in tomatoes and tip burn in lettuces. Zinc toxicity was noted first as a result of the use of galvanised pipework, but many suitable alternatives are available now and this source can be easily avoided.

root leachates may be important in hydroponic systems

It has been known for many years that considerable quantities of organic compounds are released by plant roots into the soil where they are either utilised by soil organisms or leached out by rainfall. This phenomenon presumably occurs in hydroponically grown plants, but it does not appear to have received much attention. It is of interest for two reasons. On the one hand, root leachates would presumably accumulate in closed hydroponic systems and might be expected to affect the balance of nutrients. On the other, if some way could be devised to reduce or prevent the leaching out, significant effects on yield might be achieved. The prevention of the leaching out of compounds in soil grown crops might be expected to have a detrimental effect on soil organisms and, indirectly, on plant growth. This should not be the case in hydroponically grown crops and this topic would be an interesting one for research experiments.

Let us test your understanding of some aspects of hydroponics before moving on.

SAQ 6.9

Which of the following are advantages of hydroponics? Give reasons for your decision.

1) Can be used on derelict land.

2) Uses water and minerals more efficiently than OFA.

3) Produces crops more cheaply than OFA.

4) Crops can be grown largely independently of outside temperature.

5) Uses less land than OFA.

6) Allows control of nutrient composition applied to the roots.

Further information comparing hydroponics with OFA is given in Table 8.3, which reports typical yields for crops grown hydroponically in the southwest region of the USA with typical 'good yields' for OFA.

Crop	Hydroponic CEA			OFA2
	Yield/crop (tonne ha^{-1})	No. crops/ year	Total yield (tonne ha^{-1} year^{-1})	Total yield (tonne ha^{-1} year^{-1})
broccoli	32.5	3	97.5	10.5
bush beans	11.5	4	46.0	6.0
cabbage	57.5	3	172.5	30.0
Chinese cabbage	50.0	4	200.0	-
cucumber	250	3	750.0	30.0
eggplant	28.0	2	56.0	20.0
lettuce	31.3	10	313.0	52.0
pepper	32.0	3	96.0	16.0
tomato	187.5	2	375.0	100.0

Table 6.3 Comparison of production by hydroponics and OFA. Source M H, Jensen and W L Collins (1985). Hydroponic vegetable production. Horticultural Reviews 7, 483-558.

Examine Table 6.3. To what can we escribe the greater productivity of hydroponics as compared with OFA, utilising the information provided in the table?

With the exception of lettuce, the greater productivity is accounted for by a higher yield per crop of the hydroponically grown crops compared with the total yield of that crop grown in OFA, coupled with the fact that several crops can be produced per year in hydroponics.

6.11 Pests and diseases can be a problem

greenhouse
crops are not
immune to
pests and
diseases

There is a common assumption that plants in greenhouses can be more easily kept free of pests and disease than plants in OFA because they are enclosed. This is not true. Pests and diseases can be introduced either on the clothing of operators or simply by gaining entry through doors and windows. Pest populations can spread with alarming speed inside the greenhouse irrespective of whether soil or hydroponic systems are in use. Hydroponics has the added problem of the quick spread of root diseases in closed systems. That having been said, the enclosure of the growing area should make it easier to tackle problems than in OFA.

6.11.1 Root diseases

Tomato roots are most susceptible to pathogen attack when the first fruits are formed and are beginning to swell.

SAQ 6.10

What is happening, physiologically, in the root when the first fruits are forming, which might offer a clue to this vulnerability?

restricted use
of pesticides

Common pathogens of tomato roots include *Fusarium, Verticillium, Pythium* and *Phytophthora spp*. Tomato varieties vary in their resistance to these pathogens and this is an important character in breeding programmes. A number of pesticides are available for their control but these are not always cleared for use in greenhouses. This is particularly the case in the USA where no chemicals have been cleared for greenhouse use by the Federal Drug Agency.

A bacterial canker has been reported as being a particular problem in NFT systems but it transpires that this is primarily a seed-borne disease and seeds can be surface cleaned more easily than any other plant part. Thus good management practices enable this problem to be avoided.

6.11.2 Foliar diseases and pests

There are no foliar diseases or pests associated exclusively with hydroponics so the measures developed for OFA can be applied here, providing testing agencies have cleared them for greenhouse use. The reason why this has not been done in the USA is because of the small size of the market. Chemical companies will not pay for tests unless there is a likelihood of a good return on the investment.

6.11.3 Integrated pest management

pesticide
resistance and
biological
contact

integrated pest
management
(IPM)

The problems of clearance of chemicals notwithstanding, an alarming phenomenon has arrived on the scene *vis* the development of resistance to pesticides by many pathogens. This has revived worldwide interest in the use of biological control - the deliberate introduction of natural enemies of pests. This was first introduced in the 1920s in England with the use of the parasite *Encarsia formosa* to counter whitefly infestations of tomato, but its use was discontinued because of the development of chemical pesticides. This method has been resurrected and several others added for example, to control red spider mite. Although superficially very attractive, these methods are rarely 100% effective. For this reason growers are now encouraged to practice what is called Integrated Pest Management (IPM). There is no standard definition of IPM but it generally consists of a carefully structured combination of biological controls, plant breeding, cultural practices and chemical control. Cultural practices include plant spacing, ventilation and irrigation methods as well as general hygiene, the idea being to minimise the likelihood of attack. Chemical control is used in IPM only as a last resort and then in the most efficient way possible.

6.12 Which system of hydroponics is the best?

capital outlay, operational costs and yields are the main influences on choice

The hydroponic systems described vary one to another regarding the capital outlay required, the day to day running costs and the yields of crops obtained. Capital costs vary enormously from one country to another but it is generally agreed that, if sand is freely available, it is the cheapest system as far as setting up costs are concerned. This is followed, in order, by bag culture, rockwool, NFT, and the raceway (floating) system. Some comparisons of yield have been carried and, although they are perhaps not as numerous as one might like, they suggest that the various systems are fairly comparable to each other in yield. This is the conclusion of Jill Vaughan, a soil scientist working for MAFF in the UK, in an article in The Grower, July 13, 1989. Whereas NFT systems predominated in the UK during the early 1970s, there has been a gradual replacement of it in many areas by aggregate systems, particularly based on rockwool. Thus in England in 1989 NFT accounted for only 11% of the total area of protected soilless culture and rockwool 85%, the remainder being perlite. This reflects the fact that NFT, in its usual closed mode of operation, is technically more demanding than open aggregate systems. Despite these points, individual growers will always find a system that suits them best.

rockwool is currently favoured

6.13 Controlled-Environment Agriculture-Phytofarming

controlled-environment agriculture may greatly increase yields

Modern glasshouses have very narrow supports for the glass and thus only minimally interfere with light penetration. Nevertheless, no glasshouse can make a dark day bright or a short day long. As we saw in the previous chapter there are also problems of overheating. These problems are avoided by the use of controlled environments, in which temperature and humidity are routinely controlled and light is provided totally by artificial sources. At least one company in the USA now grows vegetables in controlled environments, the plants being supplied with nutrients and water by hydroponics. The system is a closed one and uses nutrients flowing along troughs, recycling and replenishment being carried out as in NFT systems, but using a deeper mass of nutrient than that used in NFT. Plants such as lettuce are grown on expandable racks which allow for plant growth by enlarging the growing area as the plants mature. The company which developed this system has developed novel, award winning packaging systems and they have coined the term phytofarming for their type of operation. Published figures (Table 6.4) suggest that phytofarming yields are about 100 times that of field production.

	Relative Yield
field	1
greenhouse/soil	10
greenhouse/hydroponics	12.5
phytofarming	100

Table 6.4 Comparison of annual yields of lettuce. Source N Davis (1985) Food Technology 39 (10) pp 124-126 and 134.

6.14 Future applications

It is claimed by those already active in hydroponic vegetable culture that it will become increasingly important due to its greater productivity than other systems. The continued expansion of towns and villages, leading to a continued reduction in the amount of arable land, will contribute to this. In the final analysis, however, cost will always remain the major factor and this limits the expansion of hydroponic growing systems, especially those at the technically more advanced end of the process. The costs of power are a major factor in the financial viability of such systems. Power is much cheaper in the USA than in Europe and it is striking that no commercially viable equivalent to phytofarming has been developed outside the USA. The fact that much of Europe's oil is imported means that it is not in total control of power costs. For this reason, this author considers that the development of hydroponics in Europe will stop short of the phytofarming level of technology because of risks of sudden changes in fuel costs.

future expansion of phytofarming may depend on fuel costs

Summary and Objectives

Hydroponics involves growing plants with their roots in a solution of nutrients. The process may involve the use of purely liquid, non aggregate systems in various forms; NFT uses a thin film of recirculating fluid; floating hydroponics grows plants in polystyrene rafts floating on nutrient solutions; in aeroponics the nutrient is applied to plant roots as a mist. Hydroponics can use aggregates as a root support and common materials include sand, peat, rockwool, perlite and a number of synthetic polymers.

Liquid systems are usually closed, in that the nutrients are recirculated, but aggregate systems can also be open, in that nutrients are simply delivered to the plant and excess runs to waste. Rockwool provides a close-to-ideal root environment and is the preferred aggregate in England. The use of perlite in the UK was pioneered in Scotland and is still very popular there. Pests and diseases can spread rapidly in the protected environments used with hydroponics and extra care must be taken to avoid the introduction of infections. High technology systems, including phytofarming, are virtually independent of the external environment but their spread will depend upon the level and stability of fuel costs.

Now you have finished this chapter you should be able to:

- explain the meaning of the term hydroponics and compare it in various ways with OFA;

- describe a range of hydroponic systems and analyse designs for efficiency of production;

- describe the uses of sand, rockwool and perlite in aggregate systems;

- show an understanding of the factors affecting productivity, including nutrient content and conductivity, by analysing provided data.

Diversity of plant secondary products

Diversity of plant secondary products

7.1 Plant products of economic importance

All animal life is dependent on plants since plants trap the energy of sunlight and liberate oxygen through photosynthesis. At the same time they fix the carbon from the atmosphere which is consumed as food. The vast bulk of arable crops are indeed grown to provide food for consumption by humans or domestic animals. It is interesting to note that about 80% of the total edible dry weight produced each year is provided by only eleven species. However, Man makes use of a huge range of plants for purposes other than bulk food production.

∏   Try thinking of reasons why plants are cultivated or their products collected from wild populations. Compare your list with Table 7.1.

Products	Use
bulk food production	direct human consumption
	animal grazing
flavouring and perfumes	used as herbs and spices
crop protection	natural insecticides
medicine	whole herbs or purified chemicals
stimulants	tobacco and other drugs
construction materials	timber and thatching products
fibres	cloth and rope
industrial processes	rubber (natural)
	vegetable oils and fats
	tannins
	dyes
ornamentals	in parks and gardens and as cut flowers

Table 7.1 The diverse uses to which plants are put by man.

Many of the plants used for these purposes have traditions going back thousands of years. In many cases, trade in plants has been responsible for the spread of civilisations. The maritime explorations of the Renaissance period were partly to find easier routes to the areas of spice production in Asia. However, old traditions have been superseded by modern technology - products of the petrochemical industry now fulfil many of the roles previously played by plants.

⫨ See if you can list reasons why the study of traditional uses of plants may still be important.

The alternative products may be more expensive (particularly to developing countries) and may have a greater environmental cost. Industrial synthesis of many pharmaceutical compounds remains difficult because they must be of the correct chirality (D or L forms). In many cases the properties of plants used by people in remote regions are not widely known. Of particular concern is the loss of knowledge of medicinal plants from, for example, the Amazonian region as the established culture is destroyed through widespread clearing of the forests. Plants of worldwide importance may be lost forever.

⫨ See if you can think of examples of compromises between using natural and synthetic products?

natural products may be modified chemically or biologically

Substances extracted from plants are used as feedstocks for industry. In these processes chemical modifications may be carried out using conventional chemical engineering technology or by biotransformation (in which enzymes in cultured cells are used to change the structure of certain compounds supplied to them in the medium). Many plant-derived chemicals are used in industrial situations, eg sucrose (from cane or beet) and may be converted to a wide range of chemicals in bacterial fermentations. Also the use of isolated enzymes to mediate chemical transformation has become increasingly important. Thus there are numerous examples of compromises between using natural and synthetic compounds. These topics have been described in detail in the BIOTOL text, 'The Technological Applications of Biocatalysts'.

This chapter and the next are concerned with the production of crops other than those grown for bulk food, timber or fibre production. As we shall see, they continue to be of an importance disproportionate to the areas of land devoted to their cultivation. We will begin by providing an overview of the range of plant-derived secondary metabolites before examining some of these in detail. Do not be put off by the catalogue nature of this chapter. It is our intention to provide you with some knowledge of the great diversity of plant metabolites and the uses to which they are put.

In the next chapter, we shall discuss the factors which affect secondary product yield and the methods used to maximise product yield. We will use a number of case studies to explain how improved yields may be achieved.

7.2 Secondary metabolites

In order to live, any organism must continually perform a huge range of chemical reactions. This process is known as metabolism and may be divided into catabolism (the degradation of large molecules) and anabolism (synthesis of biomolecules). Investigations by biochemists have demonstrated the existence of a large number of metabolic pathways. Conversion of one substance to another proceeds by a number of clear steps, each catalysed by an enzyme. An example is the glycolytic pathway, part of which is shown in Figure 7.1.

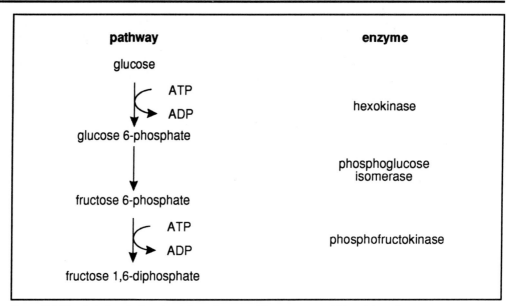

Figure 7.1 The first steps in the glycolytic pathway and the enzymes which catalyse them.

∏ What may govern the rate of the reactions shown in Figure 7.1?

feed back
inhibition

Much more detail concerning enzymology and metabolism is provided in other books in this series (see for example, 'Energy Sources for Cells' and 'Biosynthesis and Integration of Metabolism'), but in general a phenomenon known as feedback inhibition is most common. The turnover number of certain regulatory enzymes in a pathway is reduced if their products accumulate. In the example shown, phosphofructokinase is a key regulatory enzyme.

primary and
secondary
products

Glycolysis, the first stage of respiration, is a clear example of primary metabolism. It is essential for survival and is common to many organisms from bacteria to Man. Primary metabolism includes the production and use of essential sugars, amino acids, fatty acids, nucleotides and the polymers derived from them (polysaccharides, proteins, lipids and nucleic acids). In addition, most organisms also have a set of pathways which comprise their secondary metabolism. It is these secondary products, or secondary metabolites, which are used as medicines, flavourings, dyes etc as was outlined in Section 7.1. The function of secondary metabolites (also sometimes described as 'natural products') in the organisms in which they are produced is frequently unclear. In others it is obvious, such as the rendering of leaves unpalatable to prevent them being eaten by insects, or the colouring of petals to encourage pollinating insects.

∏ Is it possible for secondary products to confer no benefit to the plants which produce them?

We usually believe that all aspects of an organism must have, at some time, conferred an evolutionary benefit. However, it is possible for the need for a particular compound to have disappeared as conditions have changed. It may, however, still be synthesised if it is not a particular drain on metabolism. Another factor is plant selection and breeding by Man. Plants which are high producers have naturally been preferentially cultivated and these may have secondary products with no apparent purpose.

Secondary products are generally rather small molecules (relative to the size of proteins etc) often based on a multiple ring hydrocarbon skeleton to which atoms of other elements (eg oxygen and nitrogen) are joined. They represent a bewildering array of compounds. Some further description of these must be given now so that the names will make more sense when we discuss their properties later in the chapter. We do not anticipate that you will need to remember all of the details of their structures.

7.2.1 The range of secondary products

Non-protein amino acids

about 200 different non-protein amino acids produced by plants

About 200 types occur either free or as simple peptides in plants. Some are very similar to protein amino acids (for example canavanine is an analogue of arginine). They may have significance in influencing plant-pathogen or plant-pest interactions. Some examples are illustrated in Figure 7.2.

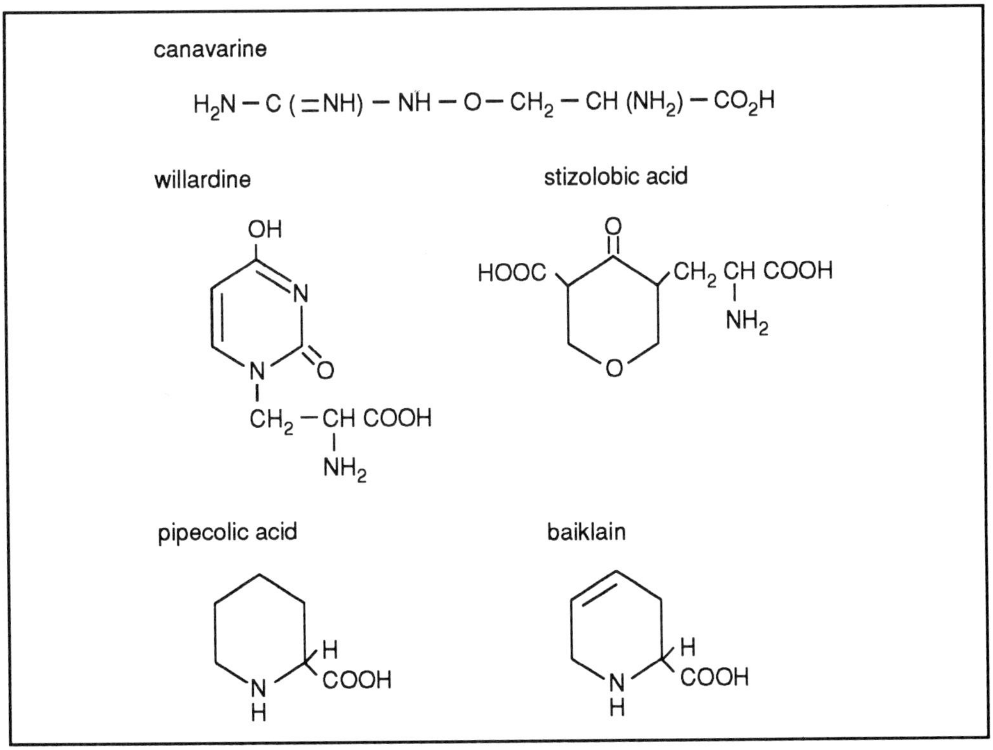

Figure 7.2 Non-protein amino acids produced by plants. Note that pipecolic acid and baiklain are 'imino' acids related to proline.

Amines

Some examples of plant amines are illustrated in Figure 7.3.

Amines may be considered as derivatives of ammonia. Polyamines (for example spermine) have been implicated in many cellular functions such as protein synthesis and cell differentiation. They may be conjugated to a range of acids (for example cinnamic acid). Amongst other functions, these may have antiviral properties. Aromatic amines also occur, for example hordenine decreases the palatability of grass. Histamine is partly responsible for the unpleasant effects of stinging nettles.

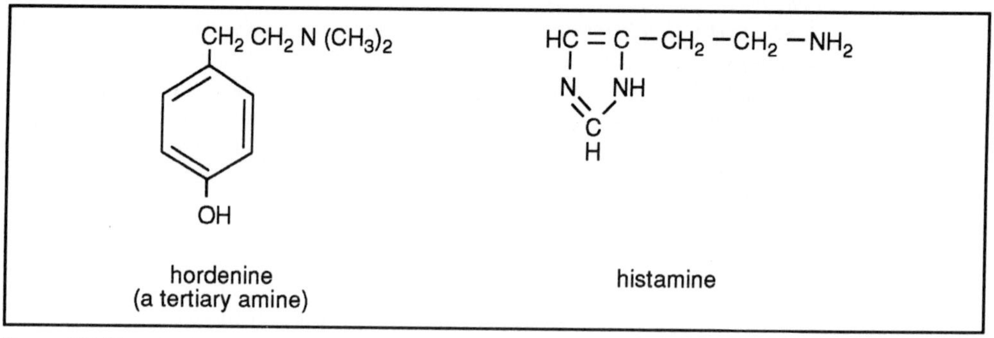

Figure 7.3 The structures of the amines hordenine and histamine.

Phenolic acids

many different
types of
phenolic acids
are produced
by plants

Phenolic acids include cinnamic acid from which many other secondary products are derived. Benzoic acid is also a pprecursor of a variety of products, including cocaine. Coumarins are mainly phenolic compounds derived from benzo-α-pyrone and are found in plants in both the free state and as glycosides (bound to a sugar molecule). Some typical plant phenolics are illustrated in Figure 7.4.

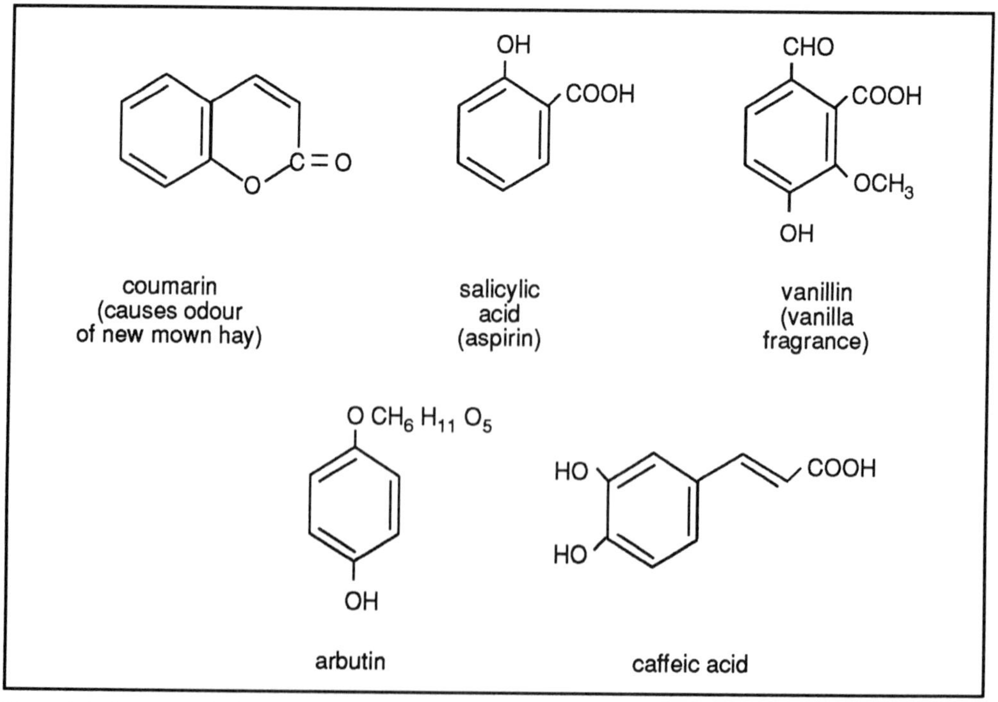

Figure 7.4 Examples of phenolic plant products.

∏ A compound we described under amines may also be regarded as a phenolic compound. Which one was it?

You should have identified hordenine (see Figure 7.3).

Alkaloids

alkaloids
embody many
sub-groups

Alkaloids are a major group of products but are rather difficult to define. They all contain a nitrogen atom and have some basicity. Usually they have heterocyclic ring structures and exhibit significant physiological activity in animals. Chemical synthesis is usually very difficult because of the D and L forms that are produced. Alkaloids may be classified according to the chemical groups of the molecules, for example solasodine (the poison in green potatoes) is a steroidal alkaloid, caffeine (in coffee and tea) is a purine alkaloid, and atropine and scopalamine are tropane alkaloids. The structures of these are illustrated in Figure 7.5.

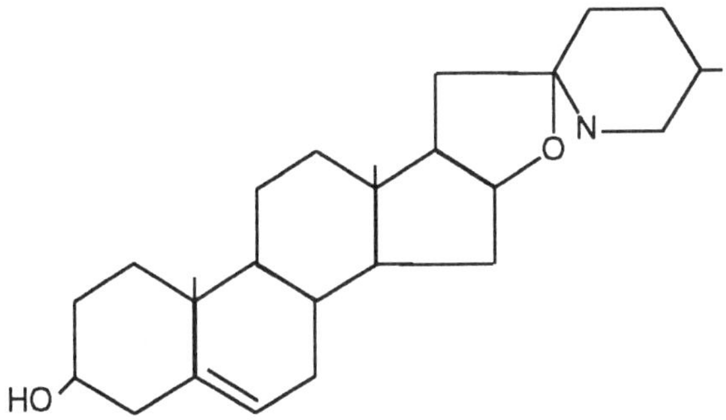

solasodine
(steroidal alkaloid)

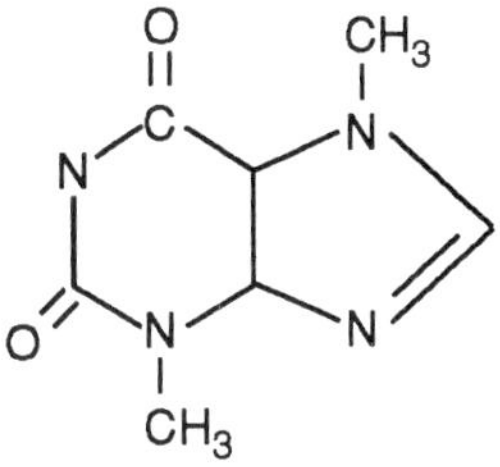

caffeine
(purine alkaloid)

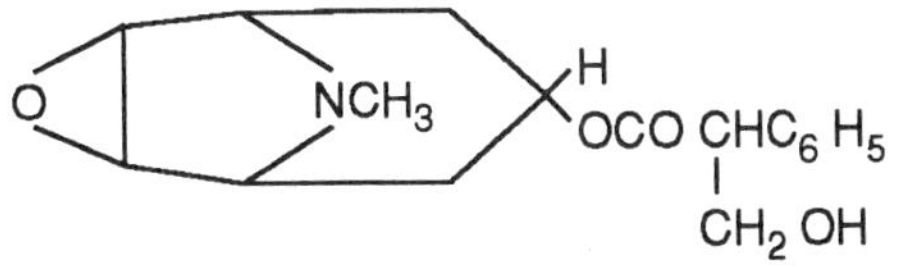

scopolamine
(tropane alkaloid)

Figure 7.5 Examples of plant alkaloids.

alkaloids
classified
according to
plants which
produce them

We can, as we have done above, group alkaloids together on the basis of their structures or on the basis of the groups of plants which produce them (for example Lupin alkaloids, *Ormosia* alkaloids, Lycopodium alkaloids etc).

∏ You may know of some plant alkaloids from your general reading. See if you can list some. (Hint: think of some drugs you know that are produced by plants).

There are many examples that could be cited, but the well-known representatives are strychnine, cocaine, nicotine, quinine and morphine.

Anthraquinones

Anthraquinones are multiple ring structures having purgative properties and are used as dyes. Good examples of these are the orange-red compounds of rhubarb stems. They are based on the following structure:

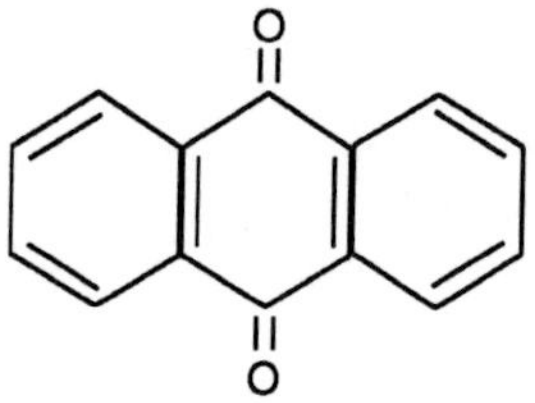

substitutions
include OH,
CH₃, CH₂OH
groups

This basic structure may contain a variety of substitutions including hydroxyl groups, methyl groups, hydroxymethyl groups and carboxyl groups. The hydroxyl groups are sometimes glycosylated, and reduced derivatives (anthranols and anthrones) also occur in some plants while others may produce dimers. There are, therefore, a great variety of structures. A few examples are illustrated in Figure 7.6.

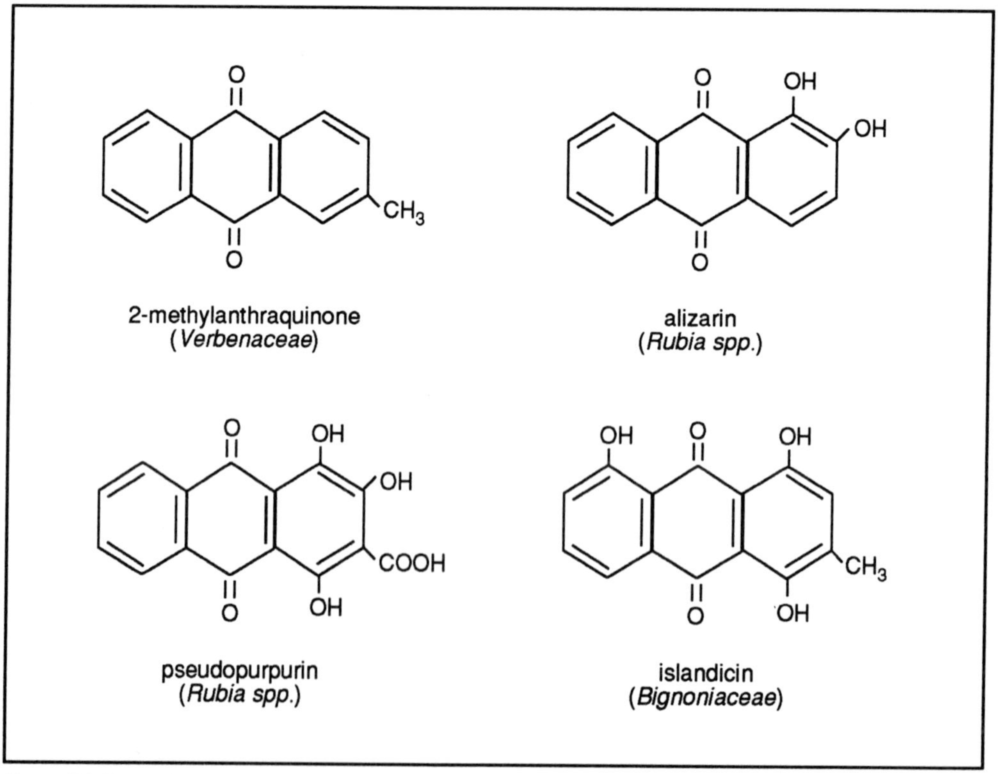

2-methylanthraquinone
(*Verbenaceae*)

alizarin
(*Rubia spp.*)

pseudopurpurin
(*Rubia spp.*)

islandicin
(*Bignoniaceae*)

Figure 7.6 Examples of anthraquinones.

Flavonoids

flavonoids are
flower pigments

Flavonoids are most easily recognised as flower pigments (orange, scarlet, crimson, purple, blue) but they occur in all parts of the plant. We can divide these compounds into four major groups on the basis of the arrangement of ring structures. Thus:

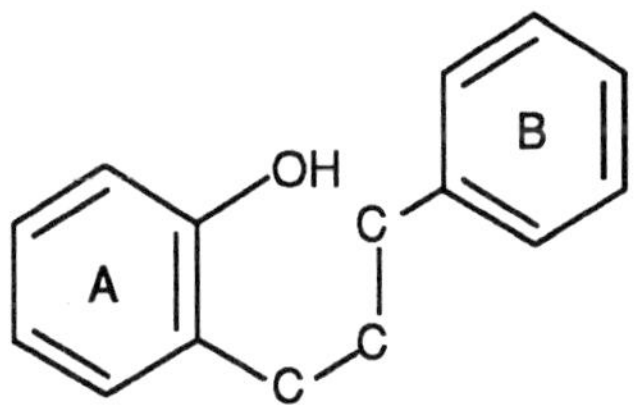

chalcones
(also dihydrochalcones)

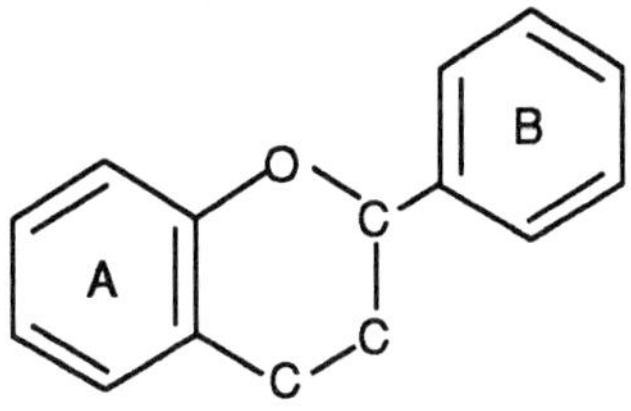

flavans
(also catechins,
flavan-3,4-diols, flavanones,
flavones, anthocyanins,
flavon-3-ols, flavan-4-on-3-ols)

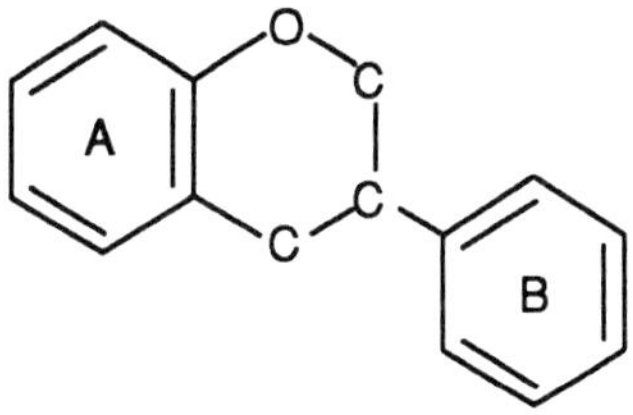

isoflavones

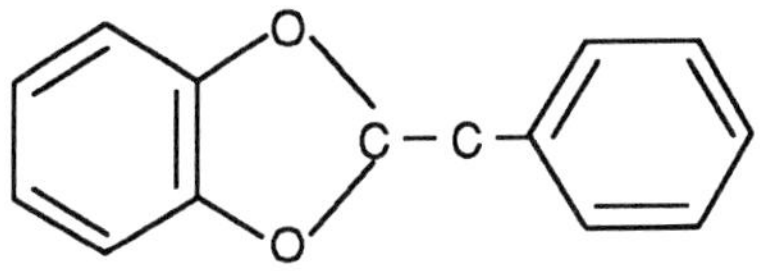

aurones

You will notice that under flavans, that we have listed other groups. These are based on the substitutions on the 3 carbons of the central ring. Thus, for example, the structures are:

flavan

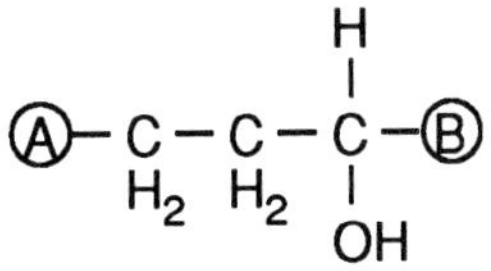

catechins

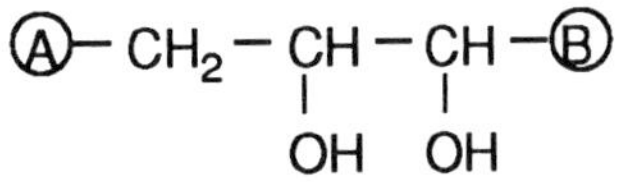

flavanones

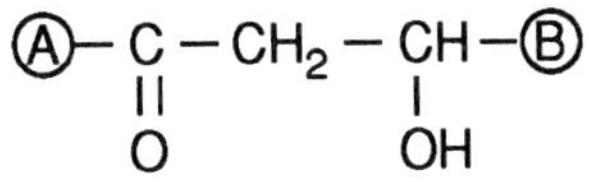

flavones

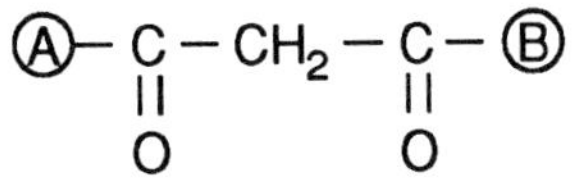

anthocyanins

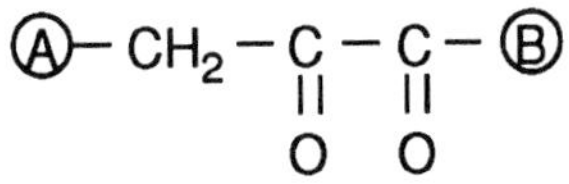

Each of these sub-groups represent a complex collection of molecules in which rings A and B are substituted. Hydroxyl groups and hydroxylmethyl groups are common, although hydroxyl groups are frequently glycosylated.

Some examples are shown in Figure 7.7. There are many hundreds of different structures based on these which have been isolated from plants.

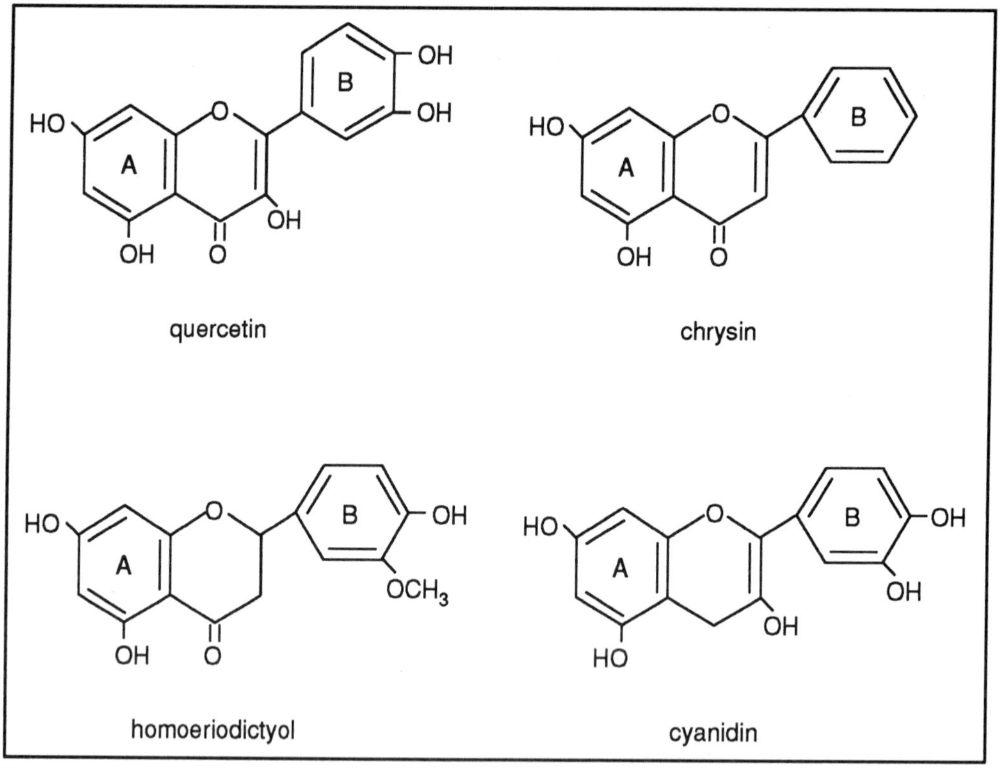

Figure 7.7 Examples of flavonoids.

∏ Examine the structures shown in Figure 7.7 and compare the configurations of substitutions in the two aromatic rings.

You should have noticed that the substitutions in ring A are in the 'meta' (2,4,6) configuration, whilst those in ring B are in the ortho/para (1,3,4) configuration. This reflects the different routes of biosynthesis of these two aromatic ring systems. Ring A is synthesised by the condensation of three acetyl groups. Diagrammatically:

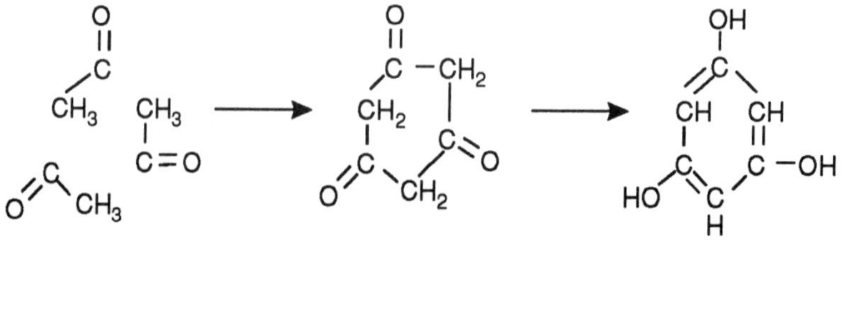

In contrast, ring B is synthesised from shikimate (a precursor of phenylalanine and tyrosine).

Next to chlorophyll, anthocyanins are the most important group of pigments visible to the naked eye. They have been extensively studied because of their economic importance for colouring fruit juice and wine.

Cyanogenic glycosides

Cyanogenic glycosides (see Figure 7.8) release hydrogen cyanide (HCN) after disruption of the tissues. This would clearly have a herbivore-deterrent effect! Cassava, a staple human food, requires careful preparation to remove all the cyanogenic compounds before it is eaten. The release of hydrogen cyanide is often dependent upon the activities of enzymes.

Glucosinolates

Glucosinolates (see Figure 7.8) are hydrophilic non-volatile compounds which are important in insect-plant interactions. They are often present in plants used as condiments for example mustard and horse radish.

$\prod$ What is the chemical nature of the $C_6H_{11}O_5$ residues of sinigrin and gluconasturtiin? (See Figure 7.8).

It is glucose (hence the name glucosinolates). We might draw the structure of these as:

$$R - C {\overset{\displaystyle \nearrow N - O - \text{sulphate}}{\searrow S - \text{glucose}}}$$

or more generally as:

$$R - C {\overset{\displaystyle \nearrow N - O - \text{sulphate}}{\searrow S - \text{sugar}}}$$

In this case, however, we should more properly refer to them as glycosinolates as the identity of the sugar residue has not been defined.

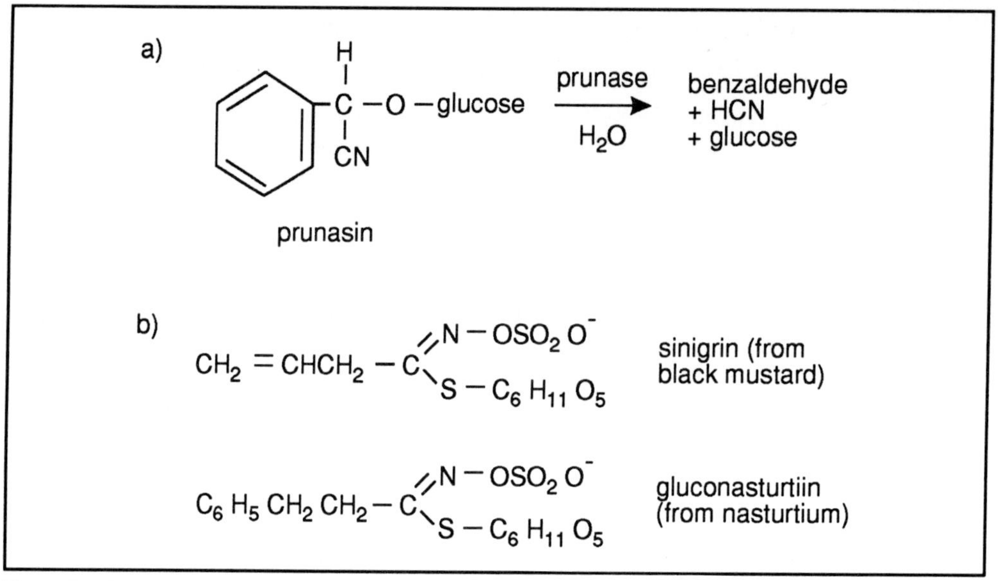
a)
H
C
O
glucose
CN
prunasin
prunase
H_2O
benzaldehyde
+ HCN
+ glucose
b)
$CH_2=CHCH_2-C$
$N-OSO_2O^-$
$S-C_6H_{11}O_5$
sinigrin (from
black mustard)
$C_6H_5CH_2CH_2-C$
$N-OSO_2O^-$
$S-C_6H_{11}O_5$
gluconasturtiin
(from nasturtium)

Figure 7.8 Examples of a) cyanogenic glycosides and b) glucosinolates. Note that HCN is released from prunasin through the action of the enzyme prunase.

Tannins

Tannins have been defined as water soluble phenolic compounds with high molecular masses (500 and 3000 Daltons) and with the ability to precipitate alkaloids and proteins.

Betalains

Betalains are nitrogenous pigments which produce colours from yellow to red in a restricted number of plants.

These are based on the structure:

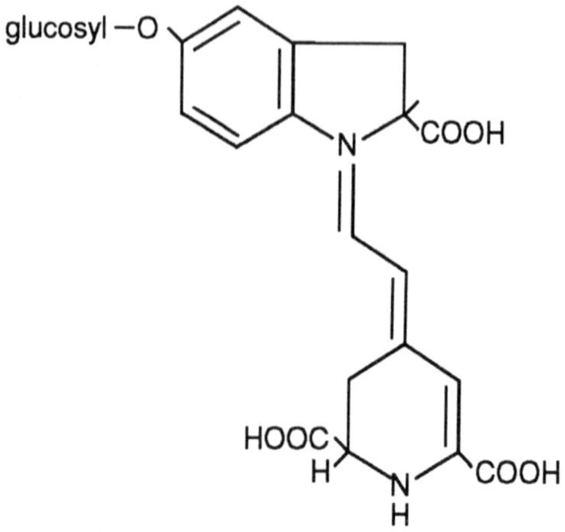
glucosyl —O
N
COOH
HOOC
H
N
H
COOH

betanin

Isoprenoids

Isoprenoid compounds are structurally based upon isoprene.

$$CH_2 = C - CH = CH_2 \quad \text{isoprene}$$
$$|$$
$$CH_3$$

∏ From your previous experience of biochemistry, you may recall several different types of compounds based upon this structure. Try to make a list of these.

The sorts of compounds we anticipate you may have listed are steroids (sterols), carotenoids and the side chains of many of the quinones/quinols that participate in electron transport systems (for example ubiquinone and menaquinone). The side chain attached to the porphyrin moiety of chlorophyll (phytol) is also composed of isoprene-like residues.

Plants produce an enormous variety of isoprenoid compounds. They can be divided up according to the number of carbon atoms they contain in the following way:

- terpenes = C_{10};

- sesquiterpenes = C_{15};

- diterpenes = C_{20};

- triterpenes = C_{30};

- tetraterpenes = C_{40}.

It is, however, more usual to divide them up according to other structural and metabolic criteria.

Here we will not give a great catalogue of these compounds, but we have illustrated the structures of some examples in Figure 7.9. You will notice from Figure 7.9 that this group includes some low molecular weight terpenes (C10), carotenoids (for example β-carotene) and a variety of structures rather reminiscent of sterols and steroids.

If you examine Figure 7.9 carefully you will realise that many of the molecules shown have many chiral C atoms, thus many isomeric forms are possible. Since various substitutions can be added to these molecules (for example hydroxylations), or methyl residues removed by oxidation (via aldehyde/keto and carboxylic acid residues), then you should realise that we are considering a very large and complex group of molecules.

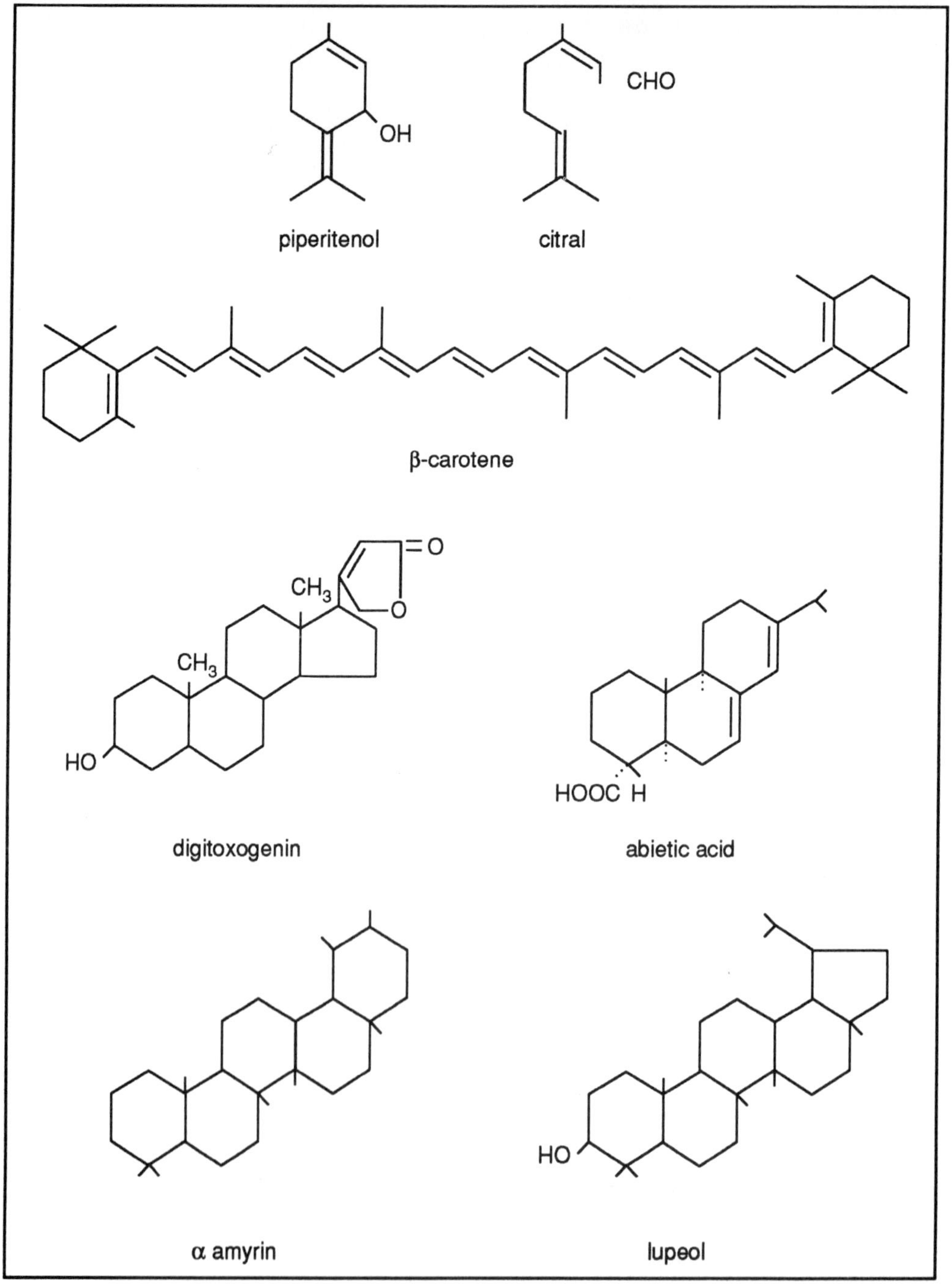

Figure 7.9 Some examples of isoprenoid derivatives produced by plants. Although these are all based on combinations of isoprene units, it is usual to consider them in separate groups. For example digitoxogenin is an example of a cardenolide. Since sugar residues are often added to such structures by glycosylation, the resulting glycosides are often called cardiac glycosides. For example, if digitoxogenin is glycosilated with the sugar digitoxose, the resulting glycoside is known as digitoxin.

Isoprenoids are responsible for many plant fragrances and colours. Terpenes are volatile compounds - they may be alcohols (eg menthol from peppermint), aldehydes (eg citronelal) or aliphatic compounds (eg zingiberine from bay). Natural rubber is a terpene polymer.

Carotenoids are responsible for most yellowish plant pigments, they contain extensive networks of single and double bonds which strongly absorb light. Lycopene, a precursor of other carotenoids, provides the colour of tomatoes, and β carotene that of carrots.

Oils

Oils are esters of fatty acids and glycerol while waxes are esters of fatty acids and higher alcohols.

Steroidal saponins

Steroidal saponins are important because they are related to mammalian hormones (and cholesterol). They are particularly common in many monocotyledonous plants (eg yams and lillies)

Cardiac glycosides

Cardiac glycosides are cardenolides containing a range of sugar varieties which have particular effects on the heart muscle (eg digitoxin - see Figure 7.9).

Carbohydrates

Besides the simple sugars such as fructose and sucrose, plants contain many large carbohydrate polymers such as starches and gums. Tragacanth and acacia gums may be used in the formulation of tablets.

Resins

Resins are solid amorphous products of a complex chemical nature including acids, alcohols, phenols, esters and inert compounds.

Π The material we have covered in Section 7.2.1 has been a little like a catalogue. We have, however, restricted ourselves to a very few examples of the enormous range of compounds produced by plants and we have made you aware of some of the major groups. Natural products chemistry is a major study in itself. We have provided some useful references in the section, 'Suggestions for Further Reading', at the end of this test if you wish to follow up this aspect in more detail. We would suggest however, that as a minimum, you should construct a large figure of your own showing the major groups of plant secondary products. Use Section 7.2.1 and Figures 7.2 - 7.9 to help you do this.

7.3 Primary and secondary metabolism

The dividing line between primary and secondary metabolism is, unfortunately, rather blurred. The two types of metabolism are strongly interlinked as shown by the examples in Figure 7.10. Thus it is often difficult to decide whether a metabolite is a primary metabolite or a secondary one.

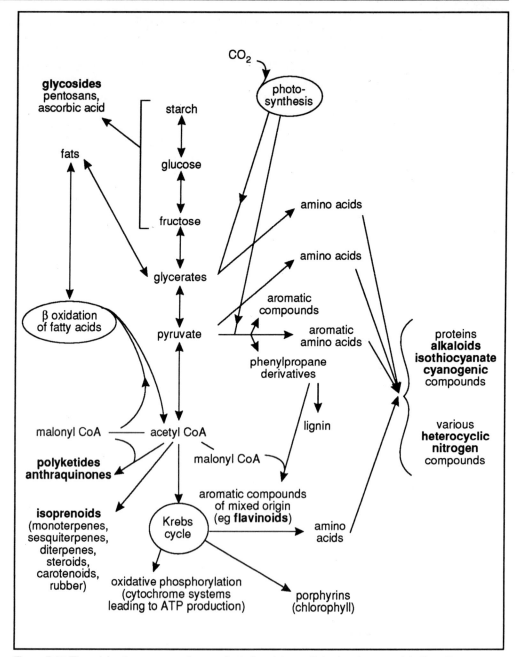

Figure 7.10 The origin of some secondary products from intermediates in primary metabolic pathways. Secondary metabolites are in bold type.

Much work has been conducted over the last 50 years to elucidate the details of primary and secondary metabolic pathways.

∏ Can you name a technique which was central to biochemists' approach to elucidating metabolic pathways?

There are many specific techniques that have been employed but the utilisation of radiolabelled tracers has been critical. A precursor containing, say, tritium (^{3}H) or carbon-14 (^{14}C) is administered to an organism or a cell extract and the identities of the biomolecules that have been produced are determined. A sequence of interconversions can thus be proposed.

∏ Examine Figure 7.10 and note that photosynthesis, glycolysis, fat metabolism, the Krebs cycle and oxidative phosphorylation are regarded as components of primary metabolism. Note that the secondary metabolites are derived from the products of primary metabolites.

SAQ 7.1

1) Which of the following would you consider primary and which secondary metabolites? There is one 'trick' item in our list.

glucose	pyrethrin
digitoxin	hexokinase
lead	vincristine
atropine	chlorophyll

2) Would you consider carotenoids primary or secondary products?

7.4 Production of secondary metabolites by plants

There are many different ways of classifying plant secondary products:

- by their function (eg drugs acting on the heart, yellow dyes etc);

- by species (eg products of *Atropa belladonna*);

- by chemical compound (eg atropine);

- by plant parts used (eg products extracted from roots or leaves).

Here we shall base our considerations on the uses of substances derived from plants.

7.4.1 Pharmaceuticals

The science of pharmacognosy (literally to acquire knowledge of a drug) is mainly concerned with naturally occurring substances which have medicinal properties and is closely related to both botany and plant chemistry.

perhaps 30% of plant species have been used for medicinal purposes

It has been estimated that at one time or another over 30% of the world's plant species have been used for medicinal purposes. Although in many cases the active ingredient(s) have not been isolated. Approximately one third of the prescription drugs in the US contain plant components. This represents consumer spending of $8000-10 000 x 10^6 annum^{-1}. So we are, indeed, discussing 'big business'.

Historically, the Chinese and Indians were the first to take full advantage of medicinal plants but the civilisations of Egypt, Greece, Rome, Arabia, Europe, Africa, Latin and North America all recorded their use. Books describing details of pharmaceutical preparations are known as pharmacopoeias.

pharmaco-
poeias

A scheme of classification based on pharmacological action emphasises pharmaceutical uses but may be extended to include those plants which are not, for varied reasons, used as drugs (eg many alkaloid and glycoside containing plants). Note that some plants (eg *Papaver* and liquorice) each contain a range of compounds falling into different pharmacological categories.

Even very large compilations of such plants are incomplete. Here, it is only intended to give an impression of the range of plants which can provide the various classes of drugs. Do not feel you have to remember them all. More details are available in books such as that by G E Trease and W C Evans, Pharmacognosy (Balliere Tindall, London). Note that some important groups of drugs, such as general anaesthetics, consist only of synthetic chemicals.

∏ We are now going to provide you with a brief overview of the pharmacological effects of plant materials. You might find it helpful to construct a summary table as you read the following sections, using the following headings. Note that we have included some examples.

Drug (or drug group)	Example(s) of plants producing the drug	Action
atropine (alkaloid)	Solanaceous plants	dilation of the pupil
cannabis	hemp	stimulates the CNS

Action on the nervous system

∏ The mammalian nervous system may be divided into two main systems. What are these?

These are the autonomic and central nervous systems (CNS), which function quite differently from each other. The former is concerned with the basic involuntary functioning of the body (eg heart beat and gut movements) while the CNS deals with co-ordinated voluntary actions. Two important neurotransmitter substances of the autonomic system are acetylcholine and noradrenaline. Chemicals which either mimic or antagonise these have a marked effect on the body. The alkaloid atropine, for example (from Solanaceous plants) is used to dilate the pupil of the eye during surgery. Other drugs include ephedrine (from *Ephedra spp.*) and ergotamine (from the *Claviceps* fungus - ergot).

plant products
acting on the
autonomic and
central
nervous
systems

Substances acting on the CNS may be broadly classified into two groups:

- compounds that stimulate mental activity (eg cannabis from hemp, caffeine and theobromine from coffee and cocoa and cocaine from *Erythroxyhim coca*);

- compounds that depress mental activity (eg reserpine from *Rauwolfia spp.*).

Morphine, the principal alkaloid of the opium poppy is used for the relief of severe pain (an analgesic drug).

Action on heart muscle

Digitalis glycosides

quinidine

Many factors affect the regulation of the heart beat but comparatively few drugs are used which act directly on heart muscle when it is failing. In the West, steroidal glycosides of *Digitalis spp.* (foxgloves) are mainly used to impart a slowing and strengthening effect. Similar drugs may be extracted from *Strophanthus spp.* and *Canvallaria majalis* (lily of the valley). Quinidine from cinchona bark (the source of quinine) is used to control atrial fibrillation.

Action on blood vessels

ergotamine and vaso-constriction

Action on blood vessels may be either to cause vasoconstriction or vasodilation. Some drugs are particularly effective on certain systems. Ergotamine causes peripheral vasoconstriction and is used to treat migraine, while papaverine (an opium alkaloid) injection is used to treat arterial embolism because it causes vasodilation.

Action on the respiratory system

ephedrine as a bronchiodilator, codeine alleviates coughing

Action on the respiratory system falls into several categories: ephedrine has a bronchiodilator effect and is used in the treatment of asthma; compounds in liquorice root and ipecacuanha act as expectorants and alkaloids such as codeine (from *Papaver*) alleviate coughing.

Action on the gastrointestinal tract

codeine and morphine reduces diarrhoea, laxative properties of some plant products

The gastrointestinal tract is subject to many ailments and a wide range of drugs are used to control them. Diarrhoea may be reduced by morphine and codeine while anthraquinones (in senna and rhubarb) and glycerides (in castor oil) have laxative properties. Derivatives of glycyrrhetinic acid from liquorice aid the treatment of peptic ulcers and atropine reduces muscle spasms. Many plant extracts (bitters) have been used to stimulate appetite (for example gentian and cinchona).

Action on the uterus

During childbirth, ergometrine has a direct stimulant action on muscle.

Muscle relaxants

tubocurarine is a muscle relaxant

Muscle relaxants are used during surgery to minimise the doses of general anaesthetics that are needed. Tubocurarine and its derivatives are most frequently used. It has its origin in curare - the arrow poisons of Brazilian hunters.

Action on the skin

antiseptics and anti-irritants

Many plant products such as oils and gums are used in the preparation of ointments, creams and lotions. They may also have a pharmacological effect in their own right for countering irritation (eg camphor) or as antiseptics (eg eucalyptus and thyme oils)

Anti-inflammatory drugs

drugs derived from diosgenin

These drugs are usually synthesised using plant steroids as starting materials (for example diosgenin and hexogenin) or steroidal alkaloids from yams or Solanaceous species. Hormones for use as oral contraceptives are produced in the same way and there is a large world market for these products. Colchicine (an alkaloid from *Colchicum autumnale* is used to reduce the swellings of gout.

Drugs against malignant diseases

vinblastine and vincristine

taxol

These drugs have been used for centuries as herbal remedies and there is currently much interest in developing these for novel cancer cures. Many appear to work by inhibiting mitosis and they are generally cytotoxic, killing both normal and cancerous cells. Selectivity may result from factors such as transport into the cell and attachment to growth regulatory macromolecules. The most established higher plant materials used in cancer therapy are the alkaloids vinblastine and vincristine. They are produced by the rosy periwinkle (*Catharanthus roseus*) and are highly purified before use. The alkaloid taxol, from the pacific yew tree, has also been shown to have antitumour properties but clinical development is limited by the supply of bark, many trees of the species have already been destroyed.

Treatment of infections

quinine

artemisinin

santonin
thymol

Fungal cultures remain the source of most antibiotics for use against bacterial infections, but plant products are used against some other disease-causing organisms. Quinine, from the bark of the Cinchona tree, was the main antimalarial drug for several hundred years before the introduction of synthetic alternatives. It is still of considerable importance. *Artemisia annua* has been known by the Chinese for 2000 years to have antimalarial properties. Artemisinin (a product of this species) has recently been developed as a purified substance. It is effective against strains of *Plasmodium* which are resistant to quinine. Another *Artemisia* species produces santonin which has antihelminthic properties (against roundworms). Thymol was once much used in hookworm treatment.

∏ Are virus diseases easy to combat?

DAP 30 and MAP 30 as a treatment for AIDS

Few drugs of any sort are effective against viruses, usually only secondary bacterial infections may be tackled with antibiotics. The spread of AIDS has led to a huge search for antiviral drugs. Some have been found, eg DAP 30 from carnations and MAP 30 from bitter melon. These also have antitumour effects and are unusual in that they are proteins. This has the advantage that the genes for them may be readily cloned by molecular biology techniques.

A vast range of plants contain substances of medicinal value yet modern medicine tends to use synthetic compounds wherever possible.

∏   List reasons why this may be so.

There are several possibilities:

- although a species may be ideal chemically it may just not be available, eg it may only exist in the wild and be impossible to cultivate;

- supplies may fluctuate in both quantity and quality depending on the weather from year to year;

- wars, political and economic 'upsets' may disrupt supplies;

- although traditional medicine made use of whole plants or crude extracts, the modern plant-derived medicines are highly refined and tested to ensure the correct dose.

SAQ 7.2

Use your summary to help you complete the table below.

Drug name	Plant of origin	Effect	Type of drug
DAP 30		antiviral	
morphine			
	pacific yew		alkaloid
		neuromuscular relaxant	
	foxglove		
		antagonises acetylcholine, stimulates CNS, dilates pupils, reduces bronchial secretions	
quinine			
ergotamine			

7.4.2 Non-pharmaceutical products derived from plants

Dyes

The most striking feature of many plants are their brightly coloured flowers. These evolved to encourage pollination by insects, but have provided the reason for many species being brought into cultivation by Man. Plants contain many coloured pigments in addition to the green chlorophyll (which traps light during photosynthesis)..

Flavonoids, isoprenoids and betalains have been extensively exploited in the past for use as dyes. Since the development of synthetic chemistry in the nineteenth century, the importance of plant dyes has decreased.

∏ Why then do we mention them?

A consideration of the chief species that were used is, nevertheless, interesting since their trade once had great political importance. Furthermore, with overproduction of food in Europe, there is scope for surplus agricultural land to be used for other crops including the production on dyes. This would decrease our dependence on the petrochemical industry.

woad The Romans found that the inhabitants of Britain daubed themselves with woad to make themselves more frightening in battle. In the middle ages, woad became the chief textile dyestuff. *Isatis tinctoria* (related to cabbage) was grown extensively in England and France and the secrets of its preparation were jealously guarded. The dye is contained in the leaves which were picked, crushed, kneaded, dried and finally fermented. Soaking cloth in a range of dye concentrations could produce any colour indigo from black to blue to green. In the 16th century indigo (*Indigofera tinctoria*) became the chief blue dye used. It was imported from Asia and later from the Americas.

saffron and extracts of dyers rocket Saffron is a yellow dye obtained from the stigmas of the crocus, *Crocus eativus*, but huge numbers of flowers produce only a tiny amount of dried product. In contrast, the whole of the above ground part of dyers rocket (*Reseda luteola*)may be used to produce a yellow pigment. The colour of a plant may change during processing. The red seed covering of annatto (*Bixa orellana*) becomes yellow as the plant parts are processed.

extracts of madder Many other plants have been used to produce various colours, eg purple or red from the roots of madder (*Rubia tinctorum*). Particularly interesting is the use of wood dyes, eg 'Brazil wood' (*Caesalpinia spp.*) and fustics, eg *Rhus cotinus*. In some cases an inorganic mordant must be used. This acts as a 'bridge' between the dye and the fabric. Different salts modify the colour produced.

Beautiful dyes may also be extracted from lichens which are symbiotic associations of algae and fungi. The supply of these is, however, largely limited to northern latitudes.

Tannins

Tanning is the process of converting animal skin into leather. Nowadays this is mainly accomplished by using chemicals such as chromium salts but traditionally, and to a limited extent today, plant extracts were used. Depending upon the plant concerned the roots, stems, fruits, bark, wood or leaves may be the most suitable material (Table 7.2).

During the tanning process, collagen chains in the skin become cross linked, protecting the hide from decay and abrasion. 'Tannins' are substances which bring this about by precipitating proteins. Plant tannins are polyphenols. They may have evolved to deter pest damage, they have an astringent taste and are involved in defence against disease. A fresh skin (hide) may absorb up to half its own weight in tannins.

Genera	Examples of tannins
Anacardiaceae	Quebracho (*Schinopsis loretzii*, *S. balansae*; heartwood)
	Sumac (*Rhus typhina*, *R. coriaria*; leaves)
	Chines (*R. semialata*; galls)
Rhizophoraceae	Mangrove (bark, various species)
Leguminosae	Wattle or mimosa (*Acacia mearnsii*; wood, bark)
	Burma cutch (*A. catechu*; wood)
	Divi-divi (*Caesalpinia coriaria*; fruit pods)
	Tara (*C. spinosa*; fruit pods), hydrolysable
	Algarobilla (*C. brevifolia*; fruit pods)
Fagaceae	Chestnut (*Castanea sp.*; wood, bark, leaves)
	Valonea (*Quercus aegilops*; acorn cups)
	Oak (*Quercus sp.*; wood, bark)
	Turkish (*Q. infectoria*; galls)
Myrtaceae	Myrtan (*Eucalyptus sp.*; wood, bark, leaves)
Rubiaceae	Gambier (*Uncaria gambier*; leaves, twigs)
Combretaceae	Myrabolans* (*Terminalia chebula*; fruit)

Table 7.2 Sources of plant tannins of commercial importance. Biochemistry of Plants, 7 (1981), Academic Press, London. * Various spellings of this word may be encountered such as myrobalans, myrabalans and myrabolams.

Plant protection

We have already seen that the production of many plant secondary products provide protection from pests and diseases.

∏ Suggest how this may be made use of in agriculture and horticulture.

Species containing plant protective chemicals may themselves be used as crops or for the breeding of new varieties. The use of varieties of plants containing high concentrations of such chemicals, however, may present problems.

∏ Can you think what these might be?

In a leafy vegetable the presence of unpleasant tasting or even poisonous compounds may be as serious a problem to humans as to potential insect pests. However, the use of plants resistant to pest and disease attack remains a very important part of plant protection since it often affords an effective means of reducing pesticide inputs.

Another possibility is to extract the active chemicals and to apply them as natural pesticides. A wide range of plant extracts have insecticidal properties. Some examples are shown in Table 7.3. Note that *Schoenocaulon officionale* (sabadilla) is also used as a rodent poison. Extracts of the seeds of *Strychnas nux-vomica* may be used to poison moles. They contain the alkaloid strychnine.

insecticidal
properties of
plant extracts

Species	Common name	Part of plant used
Annona spp.	(various)	all parts
Capsicum frutescens	chili pepper	ripe fruit
Derris spp.	derris	roots (produce rotenone*)
Allium satirum	garlic	all parts
Mammea americana	mammey	seed
Azadirachta indica	neem	seed*
Melia azedorach	Persian lilac	all parts
Chrysanthemum cinerariaefolium	insect flowers	flowers (pyrethrum*)
Quassia amara	quassia	especially the wood
Ryania speciosa	ryania	all parts
Schoenocaulon officionale	sabadilla	all parts
Acorus calamus	sweet flag	all parts
Nicotiana spp.	tobacco	leaves and stalks*
Curcuma domestica	tumeric	rhizome

Table 7.3 Examples of plants with insecticidal properties that have been exploited. The most important ones are indicated by *. Extracts may be applied as powder, smoke or liquid.

companion planting

Since some compounds which deter insect pests are volatile, companion planting may be beneficial (eg nasturtiums repel white fly when planted amongst brassicas). Mixed stands of plants are, of course, universal in nature and were once much more common in cultivated systems. Mechanisation has led to the increased use of monocultures and this has been accompanied by an increase in pest and disease problems, thus necessitating control by other means.

Non-medical drugs

tobacco products include cigarettes, pipe tobacco, cigars, snuff and chewing tobacco

The most valuable non-food crop grown in the world today is tobacco (*Nicotiana tabacum*). Tobacco contains the highly toxic alkaloid nicotine and releases a vast array of other harmful substances when it is smoked. The majority of the crop is now consumed as factory-manufactured cigarettes but other uses include pipe tobacco and cigars. Snuff and chewing tobacco are now much less important. The leaves are the parts which are used. After harvest they must be cured, fermented and matured. This takes one to two years and has great effect on the final product. Tobacco is now cultivated from Europe to New Zealand and a range of cultivars are used for different purposes.

opium poppy used for oil and narcotics

The opium poppy (*Papaver somniferum*) originated in Asia Minor and different cultivars are now used to produce an oil (edible and used for soap production) or opium. The latter is either refined for the pharmaceutical industry or sold illicitly as a narcotic.

fibre, oil and cannabis drugs from hemp

Hemp (*Cannabis sativa*) produces a useful fibre from its stem and oil from its seeds, besides the leaves and flowers providing an illicit source of hallucinogenic drugs (they contain a wide range of resinous substances).

cocaine from coca

Leaves of coca (*Erythroxylum spp.*) were chewed by South American Indians. By the 19th century the constituent alkaloids were purified and supplied as cocaine.

Many other plants and fungi have been utilised around the world for their narcotic and hallucinogenic properties eg *Ipomoea spp.* (morning glory) and *Lophophora williamsii* (a cactus).

Perfumes and cosmetics

Natural perfumes command very high prices because of consumer demand and the small quantities that can be extracted from a large amount of starting material. For example, in the production of rose oil, the volume is reduced 3000 fold by steam distillation. Although cheap substitutes may be made synthetically, they do not contain the range of compounds present in the natural products that are responsible for subtle fragrances. These are mainly due to isoprenoids and volatile oils which are often extracted by steam distillation or a process known as enfleurage (dissolving in fat) followed by partitioning into alcohol.

fragrances extracted by steam distillation or enfleurage

Perfume manufacture has a long history. In bygone days, it was often very necessary to cover up offensive odours as much as possible. Frankincense oozes from cuts in the stem of a small tree, *Boswellia carteri*, while myrrh is produced in a similar manner from *Commiphora malmal*. Flowers are extensively used. In Southern France large fields of roses, violets, jasmine, rosemary, lavender and thyme are grown specifically for this purpose. Citrus blossoms are used (for example in Eau de Cologne) and also extracts from the peel of oranges and lemons.

flowers extensively used for perfume manufacture

Fixatives are important to slow down the rate of evaporation. They, too, may be of plant origin; and are usually fragrant resins or heavy oils like clary sage and sandalwood.

fixatives used to slow evaporaton

∏ What raw materials are used in soap manufacture?

Soaps are the salts of fatty acids made by reacting lipids with alkali. While the bulk of soaps are made using relatively cheap materials such as palm oils, they are often scented by the inclusion of small amounts of more expensive products (eg rosemary and citrus oils).

use of plant extracts as cosmetic colouring agents

Plants also have more unusual functions in cosmetics, for example in certain parts of the world elaborate patterns are painted on the hands and feet using a brown dye extracted from henna leaves (*Lawsonia inermis*).

Food and drink additives

The smell, taste and colour of our food and drink is very important in determining what we are prepared to consume. In the days when food was often of uncertain quality and taste, great value was placed on spices which would make it seem more palatable. Our taste for these plants has remained to this day.

∏ What properties besides strong taste, might many spices have?

Spices, used for their flavour, frequently also have mild pharmacological properties. They are derived from a wide range of sources.

pepper

cloves
nutmeg
mace
ginger
cinnamon
vanillin

Pepper is prepared from the fruits of *Piper nigrum*, a climbing vine. They are dried and ground. Whether white or black pepper is produced depends on how ripe the fruits are allowed to become. Cloves are the dried flower buds of the tree *Suzygium caryophyllus*. They may also be distilled into an oil. The nutmeg tree (*Myristica fragrans*) supplies the spices: nutmeg (the seed) and mace (the surrounding arillus after drying). The creeping rootstock of *Zingiber officinale* produces ginger while cinnamon is the dried bark of coppiced trees of *Cinnamomum spp*. Vanilla pods are the unripe fruits of *Vanilla spp*. but have been replaced to a large extent by the synthesised chemical vanillin. This, however, does not have exactly the same flavour. Not all spices are produced in the tropics. Several are the seeds of temperate umbellifers such as caraway, dill, coriander and anise, while home grown herbs include the leaves of comfry and of the mint family.

caraway, dill
coriander

plant extracts
used to colour
food

Much food is now coloured with synthetic chemicals but natural substances can be used, eg butter and cheese may be made more yellow with annatto (*Bixa orellana*).

hops used to
flavour beers

juniper berries
used to flavour
gin

Flavourings are also added to alcoholic drinks. The taste of wine is usually derived from the fruit of which it is made. Beer however, though its bulk comes from fermented barley malt, derives much of its taste from the flowers of hops (*Humulus lupulus*). These are boiled in the wort at the mashing stage (before fermentation) and release a bitter taste together with compounds having some sedative action. Distilled spirits are also deliberately flavoured, eg gin with juniper berries. Whisky often owes much of its taste and colour to its method of storage. It may be matured in used sherry casks which release compounds both from the wood and from the remains of old wine over a period of years.

coffee tea and
cocoa

The most commonly drunk non-alcoholic drinks, coffee and tea, are both infusions of tropical crops which contain hundreds of types of molecules. These impart taste and colour and even have pharmacological action. Caffeine is an alkaloid which stimulates the CNS and has weak diuretic action. Coffee is prepared from the ground and roasted seeds of *Coffea spp*., while tea is prepared from the dried leaves of *Thea sinensis*. Chocolate is based on the seeds of *Theobroma cacao* (cocoa, not to be confused with coca which produces cocaine - see above). This also contains alkaloids, including theobromine and caffeine. All these products are mildly addictive.

A normal, balanced diet contains sufficient vitamins and minerals, but under certain circumstances supplements will be required and some of these are supplied by plants. Vitamin C (ascorbic acid) may be extracted from a wide variety of fruits (particularly citrus species). In addition, rosehips (unripe fruits) have been used as a source of vitamin A, aneurine, riboflavin and nicotinic acid. Citric acid, widely used for its antioxidant properties, is also an important product.

SAQ 7.3

Draw lines between the plants on the left and the appropriate uses on the right.

Saffron *(Crocus sativus)*	dying cloth
Derris *(Derris spp)*	tanning leather
Oak *(Quercus spp)*	insecticide production
Coca *(Erythroxylum spp)*	perfume production
Woad *(Isatistinctona)*	flavouring food
Citrus *(Citrus spp)*	illicit drug
Lavender *(Lavendula spp)*	
Cocoa *(Theobroma cacao)*	

Summary and objectives

In studying this chapter, you may have felt that it was more-or-less a catalogue. We make no apology for this. Our intention was deliberately to introduce you to the great diversity of products that can be obtained from plants and to show that there are also a multitude of uses to which such products find use. It is the great metabolic diversity of plants that brings much variety and spice (!) to our lives.

Now that you have completed this chapter you should be able to:

*   state the range of purposes for which plant products are used;

*   distinguish between primary and secondary metabolites;

*   list several types of plant secondary metabolites;

*   list several pharmaceutical products derived from plants;

*   list many plant-derived products of importance such as dyes, tannins, plant protective agents, non-medical drugs, perfumes and enhancers of food and drink;

*   appreciate the value and limitations of natural products and suggest ways that plant use and industrial production may complement one another.

Factors affecting secondary product yield

Factors affecting secondary product yield

In the previous chapter we described the range of products which can be derived from plants besides food and other bulk commodities such as fibres and timber. In this chapter we shall look in more detail at how some of these are obtained.

8.1 Yield

In Chapter 1, we defined both yield and economic yield. It should be self-evident that for a plant-derived product to be used commercially there must be a financial profit for the producer.

∏   See if you can list factors which may influence whether a financial profit may be made by producing and marketing plant secondary products.

factors which
influence
profitability

There are many factors which influence the profitability of such ventures. We anticipate that you would have included in your list such factors as the costs of producing the plants and processing these to extract the desired product(s). Also important is the amount (yield) of the desired product, the market value of the product and the size of the market.

Many factors will, of course, influence the market value of the product but ratio of supply to demand is of crucial importance. If, for example, a particular product is highly desirable and it can only be produced from plant sources then we should anticipate that the production of this product would probably be a financially viable venture. The production of pharmaceuticals from plant extracts has often been in this situation in the past because many compounds could not be produced by other means. With the advance of technology, however, many chemicals may now be synthesised. This is the case with most dyestuffs. Natural dyestuffs such as saffron (from crocus flowers) which were once tremendously important for international trade, cannot be produced as cheaply from plant sources as they can by chemical procedures. In many cases, chemical procedures enable us to make suitable alternatives to plant-derived products. Note, however, that natural dyes may still be used by amateurs whose time used in preparing the materials does not incur a labour cost.

New developments will continue to change the relative importance of natural versus synthetic chemicals. Wider economic influences may also have an effect. For example rising oil costs can make chemical synthesis more expensive while the increasing availability of agricultural land for non-food production may make natural products cheaper.

The yield of a given secondary product is a critical factor in determining the economics of its production from plants and it is this aspect that we will concentrate on.

∏   What do we mean by secondary product yield? Give suitable expressions for yield.

Yield is the amount of secondary product that is produced. Yield may be expressed in many different ways which may be more or less suitable under certain circumstances. For example:

- amount per unit area;

- amount per unit time;

- amount per unit total weight of plant produced;

- amount per unit weight of plant part containing a particular product;

- amount of purified product per unit weight of dried leaf.

The calculations in SAQ 8.1 may help to make this clearer to you.

SAQ 8.1

Suppose that a plant accumulates 2.4 kg fresh weight per m^2 of land over a three month growing period. This could be said to be its yield. However, it has a percentage dry wt of 10% and only half of this consists of leaf tissue. During the final month before harvest the leaves accumulate an alkaloid which reaches a maximum concentration of 1% of the dry weight. The plant is grown in a part of the tropics, which permits a fresh crop to be sown immediately after harvest.

Calculate the plant's yield when expressed in different ways:

Fresh weight per m^2 of land per month.

Dry weight per m^2 of land.

Dry leaf weight per m^2 of land.

Alkaloid produced per m^2 of land.

Alkaloid produced per unit total dry weight.

Alkaloid produced per unit total fresh weight.

Alkaloid produced per m^2 per month.

8.2 The choice of plants

Current estimates of the number of species of flowering plants range from 200 000 to 250 000, only a fraction of which have been chemically analysed.

∏ How, then, do we know which plants from this vast range produce useful secondary products?

Ancient Man found uses for plants empirically by trial and error. In the case of herbal remedies, this must have involved risky trials using the patient as an experimental animal. For some areas of the world such uses are well recorded but in other areas ethnobotanists are fighting against time to record information before it becomes lost due to cultural change or before the plants themselves disappear during deforestation. You should also realise that it is likely that a large number of potentially useful substances remains to be discovered. Also important is the systematic screening of plants by contemporary pharmacological laboratories.

systematic screening of plants

In the current search for new drugs (for example against AIDS or cancer), which may be made by partial synthesis from new products, a different approach is used. Investigators are faced with the systematic screening of high numbers of plants in the hope that pharmacologically active compounds can be produced from plants whose crude extracts are pharmacologically inactive and therefore unused in traditional medicine.

Most plants that have been utilised for their secondary products are from the land. They are mainly dicotyledonous although both monocotyledonous plants and gymnosperms are important as well. Algae, a large group of mainly aquatic plants, provide some substances of major importance (eg agar jelly) but have not generally been widely exploited. The sea contains a wealth of species that may become more important in the future.

∏ What symbiotic relationships between algae and fungi was once of industrial importance?

lichens

The symbiotic relationships we are looking for are the lichens. Lichens can provide many coloured dyes for cloth, but for commercial purposes this use has almost totally been replaced by synthetic chemicals. Remember that some species of fungi (eg *Penicillium spp.*) are still the source of importance antibiotics (see the BIOTOL book 'Biotechnological Innovations in Chemical Synthesis').

It must be remembered that the same secondary product can be made by several plant species. This is frequently the case with alkaloids of medicinal importance. A large group of Solanaceous plants all produce the tropane alkaloids hyoscyamine/atropine and hyoscine. These include the thornapple (*Datura spp.*), henbane (*Hyoscyamus spp.*), deadly nightshade (*Atropa belladonna*), *Duboisia spp.*, *Scopolia spp.* and mandrake (*Mandragora spp.*).

∏ Make a list of the factors which must be considered when deciding which plant to cultivates when so many species make the same products.

Some of the considerations which must be borne in mind are:

•   how well the various plants grow in a given location owing to particular climatic conditions, eg *Duboisia* is grown in Australia while *Datura* is more suitable for cultivation in Europe;

•   politics may be important. Wars and upheavals may prevent the import of important compounds from traditional areas, and then new sources must be sought, eg a shift from the use of Mexican yams for diosgenin production;

•   the pest and disease problems of particular species;

- the yield of secondary products both in absolute terms and relative to one another, for example hyoscyamine is the principal alkaloid of *Belladonna* but hyoscine is present in a higher proportion in henbane;

- other compounds which may be produced: poppies provide opium alkaloids from their latex and edible oil from their seeds, while hemp provides fibres from its stems, oil from its seed and narcotics from its leaves and flowers. However, a plant variety is usually selected on the basis that it is the best for one specific purpose.

Another important point concerning the choice of plants for secondary metabolite production is that all members of a species are not equally able to make (and accumulate) specific metabolites.

8.2.1 Plant breeding and selection

cultivars, plant breeding and genetic engineering

The division, by Man, of plant types into species is somewhat artificial since hybridisation between them is often possible and great variation exists within many species. It is this variation which has been exploited for the selection of cultivars (cultivated varieties of plants). Traditionally, selected plants have been crossed to create new combinations of genes (plant breeding) but the new molecular biological techniques now available offer exciting prospects to extend this work further. The traditional method is referred to as plant breeding while the method involving molecular manipulations is often referred to as genetic engineering.

Other books in this series explain the principles of these methods for plant improvement in much more detail (see for example 'Biotechnological Innovations in Crop Improvement', 'Techniques for Engineering Genes' and 'Strategies for Engineering Organisms'). Here we shall concentrate on their implications for secondary metabolite production.

∏ How may the genetic characteristics of a plant be changed?

The answer to the question is that there are three basic mechanisms:

- by mutation;

- by chiasmata formation during meiosis;

- by changes in the number of chromosomes.

Let us examine each of these in a little more detail.

Genetic information is carried in the DNA of the chromosomes in the nuclei of plant cells. DNA is composed of large numbers of bases strung together and it is the precise sequence of these bases which determines the inherited characteristics. Changes in the base sequence alter the genes and thus the characteristics of the plant, they are known

mutation   as mutations. These changes may occur spontaneously during cell division but they are often the result of the presence of some factor such as radiation or certain chemical mutagens. Most mutations are either deleterious (for example damaging the DNA which codes for a vital enzyme) or have no effect (because the DNA plays only a supporting role and does not contain genes at all). Occasionally, however, a mutation confers an advantage, for example a new or modified enzyme may be produced which enables the plant to produce a novel secondary metabolite which confers resistance to pest attack. A plant with such a gene may then out-compete others of its species and pass on the characteristic to the next generation. Variation also results from the rearrangement of genetic material which takes place during meiosis (the production of

chiasmata   gametes). This rearrangement occurs at chiasmata (crossing over points). Also the genetic constitution of a plant (or a plant cell) may be changed by changes in the number of chromosomes.

ploidy   The chromosomes of a cell can usually be arranged in pairs by their shape when examined under a high powered microscope. Usually cells contain two complete sets (one from each parent), such cells are termed diploid (or 2n). During meiosis the number is halved to give a set with n chromosomes; the cells obtained after meiosis are haploid. Many plants contain multiple sets of chromosomes, for example tetraploid (4n) and hexaploid (6n) genomes are common.

∏ How many chromosomes will the gametes of a tetraploid cell contain if n = 8?

A tetraploid cell contains 4n = 32 chromosomes, when this divides to form gametes the number will be halved to 16.

Polyploids may result from the multiplication of the chromosomes of a single species (autoploids) or following hybridisation between species (alloploids). The latter case is often a mechanism used to restore fertility in certain hybrids, eg *Datura ferox* x *D. stramonium*.

Let us examine an example to show how alloploidy can do this.

Consider a cross between the two species as shown in Figure 8.1. In such a situation, doubling of the chromosome number can occur spontaneously or may be induced by applying the alkaloid colchicine which inhibits mitotic spindle formation; chromosomes continue to divide but do not separate so the chromosome number doubles.

∏ Where does colchicine come from?

Colchicine is produced by the crocus *Colchicum autumnale*. Despite its effects on mitosis it is used, with care, as a cure for gout.

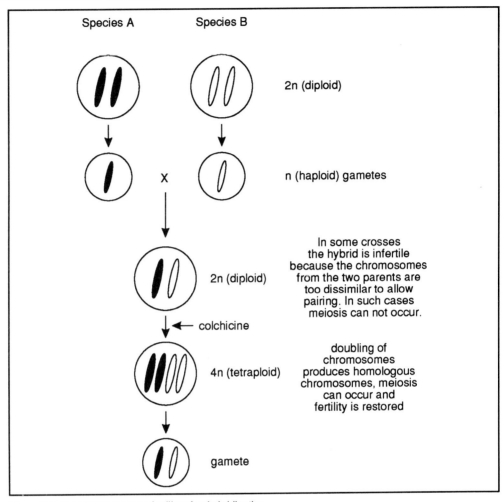

Species A   Species B

2n (diploid)

n (haploid) gametes

X

2n (diploid)

In some crosses
the hybrid is infertile
because the chromosomes
from the two parents are
too dissimilar to allow
pairing. In such cases
meiosis can not occur.

colchicine

4n (tetraploid)

doubling of
chromosomes
produces homologous
chromosomes, meiosis
can occur and
fertility is restored

gamete

Figure 8.1 Polyploidy restores fertility after hybridisation.

features of
polyploidy

Polyploid plants frequently have larger flowers, pollen grains and stomata than diploids. The influence of ploidy levels on the secondary product content of a range of plants is shown in Table 8.1. Also included in this table are data relating to haploid (n) and triploid (3n) plants. We will not explain here how such plants are produced. If you wish to follow this up, we have provided some useful references in the section 'Suggestions for Further Reading' at the end of this text.

Π   Examine Table 8.1 carefully and note the range of secondary products produced. Also note the concentration range (reported mainly as % dry weight) and the effect of ploidy. Is there a consistent pattern?

The answer is that generally tetraploids (4n) give greater yields than diploids (2n). There are however some notable exceptions (see for example *Mentha spicata*). In other words we cannot be certain that converting a diploid to a tetraploid will result in increased yield.

Plant	Constituents	Form		
		2n	4n	Others
Atropa belladonna	total tropane alkaloids, dry weight (%)		increase of about 68% over 2n	
Datura innoxia	hyoscine, dry weight (%)	0.21		0.14, 0.11 (1n)
	atropine dry weight (%)	0.03		0.01, 0.01 (1n)
Datura stramonium	total tropane alkaloids, dry weight (%)		increases of about 60-150% over 2n	
Hyoscyamus niger	total tropane alkaloids, dry weight (%)			mainly 8n increase of about 34% over 2n
Cinchona succirubra	quinine, dry weight (%)	0.53	1.12	0.27 (1n)
Opium poppy	morphine, yield/unit area			increases of up to 100%, 3n plants especially high
Lobelia inflata	alkaloid content dry weight (%) per plant	0.25	0.32-0.46 52-152% that of 2n	
Acorus calamus	volatile oil content (%)	2.1 (light oil, no detectable β-asarone)	6.8 (yellow-brown viscous oil, 2-8% β-asarone)	3.1 (3n) (yellow oil, 0.3% β-asarone)
Carum carvi	volatile oil content (%)	6.0	10.0	
Mentha spicata	volatile oil content (%)	0.48	0.05	
Digitalis purpurea	total glycosides (%)		lower or same as in 2n	
Digitalis lanata	total glycosides (%)		lower or same as in 2n relatively high content of lanatosides A and B	3.0 (3n)
Urginea indica	proscillaridin A and scillaren (%)	0.004-0.26	0.02-0.45	0.04-0.07 (3n)
Capsicum sp.	ascorbic acid (%)	0.04-0.009	0.04-0.15	
Cannabis sativa	marijuana-like activity (toxicity to fish)	1.4	2.6	

Table 8.1 The influence of chromosome number on the constituents of medicinal plants. G E Trease and W C Edwards (1983), Pharmacognosy, 12 ed Bailliere Tindall, London.

∏ The data in Table 8.1 describes the concentration of the active constituents in terms of % dry weight. Is this a good indicator of the yield of these products?

The answer is that generally the concentration of the product expressed as a % dry weight will be a good indicator of yield. The higher the % dry weight the greater the yield. However, this is not always the case. Treatments such as alloploidy may actually reduce the size (weight) of individual plants. Thus, although the concentration of the product as % dry weight of plant material increases, the total amount of product per plant may be lower. It is important, therefore, in the production of new varieties that the concentration of product-made is not the only criterion that needs to be considered. We must also consider the effects the new genetic combination will have on other properties of the plant. The ideal result from changing ploidal levels or by producing mutants would be the production of varieties that showed good growth characteristics. A good example is that provided by caraway. The generation of tetraploids from diploids resulted in a doubling of the absolute yields. At the same time the tetraploids became more frost resistant than the diploids. Furthermore the tetraploid plants can be grown as perennials whereas the diploid plants are cultivated as biennials.

aneuploids Instead of producing plants with whole sets of duplicate chromosomes, it is possible to produce plants which have additional copies of one or a few chromosomes. These are called aneuploids and the condition is called aneuploidy.

Thus, a diploid carrying an additional copy of one of its chromosomes may be designated as 2n + 1 while a diploid carrying additional copies of two of its chromosomes may be designated as 2n + 2.

Some varieties of *Datura stramonium* have this characteristic. Both increases and decreases in alkaloid contents are possible as a result.

∏ Are seedlings raised from aneuploids likely to resemble the parent plant?

Aneuploids can occur spontaneously in a population but plants will not breed true because the extra chromosomes do not segregate predictably during meiosis. Consequently, aneuploid plants with desirable characteristics must be propagated vegetatively.

So far we have been concerned only with gross differences in chromosome numbers. Many variations between plants have more subtle causes.

chemodemes Chemodemes (or chemical races) have been defined as chemically distinct populations of plants that are identical in appearance but which have different genotypes and vary in their chemical constituents. Before the existence of such races can be confirmed it must be ensured that these differences are really due to genetic factors. We shall learn later that the environment can have a profound effect on secondary metabolite production. Therefore a survey of plants growing in different situations would not confirm that differences were due to genetic differences. To test whether or not chemical differences between plants were due to genetic differences and not to environmental factors, all the plants have to be cultivated as near identical condition as possible (see Chapter 1).

Clearly the cultivation of high yielding races is desirable and there are several examples where chemodemes occur. These include *Duboisia spp.*, opium poppies, *Strophanthus spp.* (which produce steroidal saponins) and *Withania somniferum* (which produces steroidal lactones and other compounds). A comparison of the performance of selected and unselected strains of foxglove (*Digitalis lanata*) is illustrated in Figure 8.2.

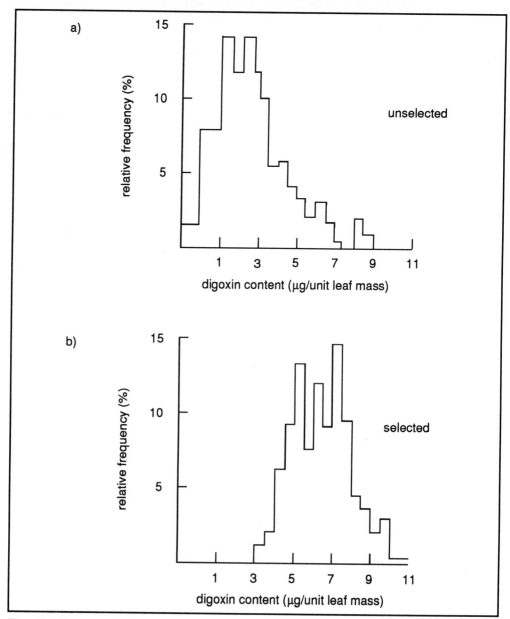

Figure 8.2 Frequency distribution of *Digitalis lanata* plants with particular digoxin contents. a) An unselected population. b) A population bred for several generations from plants with 5 or more µg digoxin per unit leaf mass.

Π Can the yield of a secondary metabolite be raised above the maximum in the original population by the method described in the legend of Figure 8.2?

The answer is no. All that is really being achieved is the elimination of the more unproductive plants so that the mean yield is raised.

We have already mentioned the phenomenon of mutagenesis. This can be used in an attempt to improve yields but the high yielding variant must still be selected out. Despite this difficulty there are numerous successful examples of this technique. Some are given in Table 8.2.

Species	Product	Mutagen
Capsicum annuum	capsaicin	sodium azide ethylmethanesulphonate
Lupinus digitus	bitter alkaloids	ethylmethanesulphonate
Papaver somniferum	morphine	^{60}Co
Datura spp.	various tropane alkaloids	X-rays

Table 8.2 Plants for which exposure to mutagens has been used for altering secondary product yield.

Note that in the case of *Lupinus digitus* (lupins) the aim was to reduce the alkaloid yield to make it suitable as a fodder crop! Sometimes the improvement in metabolite yield may be gained in a rather indirect manner, for example irradiation of peppermint produced strains resistant to *Verticillium* wilt disease to which the original plants were prone. A better yield of oil was consequently obtained. Of course, seeds from mutated plants cannot be marketed until they have been shown to breed true.

homozygosity and hybridisation

Plant breeding has traditionally been used for hundreds of years for the improvement of crop plants. This offers the means of combining, in a single variety, the desirable characteristics of several others. A common approach is to induce, by repeated self pollination, a high degree of homozygosity in two plant lines which are then crossed (hybridised). Because of the immense amount of work entailed in this approach, it is largely restricted to major food plants. More common in the field of secondary metabolite production is the hybridisation of species. This has already been mentioned with regard to chromosome numbers. Many secondary products are produced from natural and artificial hybrids. By natural hybrids, we mean hybrid strains that arise through natural processes. The production of artificial hybrids implies that the hybridisation will only occur as a result of the intervention of Man. Perhaps surprising is the fact that *Nicotiana tabacum* (the source of tobacco) is regarded as having originated from at least two parent species.

Recent developments in biotechnology offer the opportunity to improve plant varieties. A number of techniques are the key to these developments, especially:

• micropropagation;

• protoplast fusion;

• genetic engineering.

We will not examine these techniques in detail here, but we will outline their potential for improvement of secondary metabolite yields. The details of these techniques are described in the BIOTOL text, '*In Vitro* Cultivation of Plant Cells'. A wider description of the applications of these techniques to crop production are given in the BIOTOL text, 'Biotechnological Innovations in Crop Improvement.

Micropropagation

rapid
propagation

cryopreservation

Micropropagation enables plants to be multiplied vegetatively very rapidly under aseptic conditions using special nutrient media. This is particularly advantageous for high yielding hybrids which will not breed true and must be propagated vegetatively, for example clones of *Digitalis lanata* (for cardiac glycosides) and *Chamomilla recutita* (for essential oils). It is now also possible to store plant tissues frozen (cryopreservation), in which state plants can be kept safe for use in later breeding programmes. Some rare plants of medicinal value have actually been saved from extinction by the use of micropropagation techniques. Whilst this technique does not itself impose genetic variation, it does allow the speedy propagation from specimens modified by other means.

Protoplasts

Protoplasts may be made from plant cells by removing their walls using enzymes. They can then be induced to fuse with others of the same or of a different species. Whole plants, with novel genetic constitutions can then be regenerated from the hybrids and propagated by micropropagation. This technique offers some intriguing alternatives to sexual hybridisations but many technical difficulties remain unsolved.

Genetic engineering

Genetic engineering is a broadly used term, describing the construction of novel genomes. It is most commonly applied to the use of recombinant DNA technology to construct new genes or new gene arrangements and the insertion of these constructs into recipient cells. There are many techniques that may be used to create new genes (for example site-directed mutagenesis) and new gene arrangements, and these procedures are described in the BIOTOL text 'Techniques for Engineering Genes'. A number of procedures and carrier (vector) systems are available to transfer these constructs into recipient plant cells. These include both direct DNA transfer, using biolistics or electroporation, and Ti plasmids derived from strains of *Agrobacterium* (eg *A. tumefaciens* and *A. rhizogenes*). Details of these are given in the BIOTOL text, 'Strategies for Engineering Organisms'.

The creation of transgenic plants in such a manner has been widely reported for increasing herbicide and insect resistance.

∏ Why might it be more difficult using genetic manipulation, to modify the synthesis of, say, atropine than that of a seed storage protein?

Unfortunately, most secondary products are synthesised via lengthy pathways involving many enzymes, each often under strict developmental and tissue specific control. In order to improve the yield of a secondary metabolite, we would therefore need to modify the regulation of many genes. In other words, to improve yields, we may have to carry out many different genetic modifications. In contrast, a specific seed storage protein will be encoded by a single gene. Although the expression of this gene may be under developmental control (that is, it is only expressed in the seed), we would only need to replace a single regulatory sequence in order to modify the production of such a seed storage protein. Of course, even with the complex pathways leading to the synthesis of secondary products, if these are controlled by a single key enzyme, modification of the regulation of the gene coding for this enzyme may greatly influence the amount of secondary products synthesised.

Much fundamental biological work remains to be done before the yield of secondary products may be increased by genetic engineering although some success has been achieved in elucidating the regulatory enzymes for important substances such as vinblastine.

transfer of
biosynthetic
capability to
other
organisms

In the future it may well be considered easier (and more economic) to remove the genes for the synthesis of certain secondary products from plants altogether and put them into micro-organisms which can be readily cultured in bulk under industrial fermentation conditions. Plant cells too, however, can be grown in bioreactors but we shall consider the benefits and problems of this approach towards the end of the next section.

∏ Here is a list of four important methods of improving plants grown for their secondary metabolite production. For each, use the information given in the text to write down as many plants as you can for which you know this technique has been used. This exercise will help you to remember some of the specific examples.

Mutation.

Induction of polyploidy.

Interspecific hybridisation.

Selection and cultivation of high yielding races.

In doing this, cover up the list shown below.

Some species that these techniques of improvement have been used for are:

Mutation:	*Capsicum annum*
	Papaver somniferum
	Datura spp.
Induction of polyploidy:	*Atropa belladonna*
	Datura spp.
	Hyoscyamus niger
	Chichona succirubra
	Papaver somniferum
	Carum carvii
	Digitalis spp.
	Cannabis sativa
	Mentha spp.
	Urginea indica
Interspecific hybridisation:	*Datura spp.*
	Mentha spp.
	Duboisa spp.
Selection and cultivation of high yielding races:	*Digitalis spp.*
	Duboisia spp.
	Papaver somniferum
	Strophanthus spp.
	Withania somniferum

8.3 Plant husbandry techniques

collection of
wild plants

Traditionally, many plants used for dyes and medicines etc were not cultivated on agricultural land but were collected from wild populations. Some secondary products are still obtained in this way (eg *Rauwolfia* alkaloids and emetine and cephaeline - ipecacuanha - from *Cephaelis spp.*). Secondary metabolites obtained from wild populations are usually substances required only in small quantities or products obtained from plants that are very difficult to cultivate.

∏ Make a list of the reasons why it is advantageous to cultivate plants rather than to collect them from the wild.

There are a number of reasons why this is so. Here we cite some of the important ones. These are:

- greater continuity of supply is assured;

- there is greater control over the growth of the plants used. When collecting wild plants, there may be difficulty in choosing even the right species. Cultivation permits the selection of high yielding strains;

- plants may develop better under cultivated conditions because of the absence of competition. Techniques such as pruning, fertilising and pest control may be employed to improve yields;

- better facilities for treatment after harvesting may be provided. Rapid drying of leaves may be important and can be difficult to achieve following collection of plants in the wild.

For successful growth under field cultivation, it is generally desirable to mimic the conditions under which the plant flourishes in the wild. There are, however, circumstances under which manipulation of the environment can be used to increase yield.

8.3.1 Yield and the environment

∏ List the major environmental factors which you think may affect plant growth.

In part, this should be a revision of the subject material covered in Chapter 3, so here we will give a brief summary of the factors affecting yield, and provide you with a number of specific examples.

daily and
seasonal
temperatures

Both daily and seasonal temperature ranges are important. Many tropical plants will grow in temperate climates during the summer but die in the frosts of winter. Even daily temperature fluctuations can be considerable. There are optimum temperatures for the formation of many secondary products, for example more nicotine accumulates in *N. rustica* at 20°C than at 12 or 30°C.

altitude

Altitude is clearly related to temperature. There is, on average, a 1°C fall for every 200m of elevation. *Cinchona succirubra*, normally a highland plant, grows well at low altitudes but produces scarcely any alkaloids at these lower levels. The yield of *Lobelia* alkaloids, however, decreases with altitude. These types of effects of altitude have been employed to improve yields. In Ecuador, for example, vegetative growth of pyrethrum is obtained in lowland irrigated farms and the plants are moved to the highlands for flowering.

rainfall   The effect of rainfall depends on its yearly total and seasonal distribution and is modified by the temperature (related to the relative humidity) and the water holding characteristics of the soil. Very heavy rain can wash out water soluble substances even from the leaves.

day length and   Light is clearly required for photosynthesis but has many other effects on the growth of irradiance   plants (photomorphogenesis). These may depend on the day length, irradiance and spectral quality. Secondary metabolite production is also affected by light. Bright light enhances alkaloid production by *Atropa* and *Datura* and day length influences the components of volatile peppermint oils.

soil types, pH,   You are aware that soils are complex with many physical, chemical and biological minerals   characteristics. Plants vary enormously in their requirements. Soil type (sandy, silty, clayey or peaty) and pH are often the most important criteria. Also, the effect of mineral nutrition on plants has been shown to be very important.

Π   Write down the mineral nutritional requirements of plants.

Besides hydrogen, carbon and oxygen (obtained from water and air) all plants require six 'macro-elements' in relatively large amounts: nitrogen, phosphorus, potassium, magnesium, calcium and sulphur. The first three of these are especially important since they are often in short supply in soils. In addition, about seven more 'micro-elements' are essential but needed in much smaller quantities. These are iron, molybdenum, boron, zinc, manganese, chlorine and copper. While they may increase the size of plants, fertiliser applications do not always raise the secondary product yield per unit area. However, nitrogen fertilisers do often have a beneficial effect on the production of alkaloids (for example, morphine production by poppies) and this may be because these molecules actually contain nitrogen.

8.3.2 Plant growth regulator applications

The growth and development of plants is regulated by five main groups of chemicals.

Π   See if you can list these five plant growth regulator groups?

Auxins, cytokinins, gibberellins, abscisic acid and ethylene are produced ubiquitously in higher plants and effectively control, at low concentrations, cell division, enlargement and differentiation. They can also have subtle effects on secondary metabolism and have been used in attempts to increase product yield.

auxins generally reduce secondary products

Indole-3-acetic acid (IAA) is the most important naturally occurring auxin. Its main effects are on cell division and elongation and root production. Many synthetic analogues of IAA exist (eg naphthylacetic acid [NAA], 2,4-dichlorophenoxyacetic acid [2, 4-D]) and are used as herbicides and hormone rooting powders. In the plants so far investigated, applications generally reduce secondary metabolite production although NAA can increase the oil yield of *Mentha peperita*.

cytokinins generally reduce secondary products

Cytokinins principally effect cell division and inhibit senescence. Most naturally occurring cytokinins are adenine derivatives (eg zeatin). This is expensive and synthetic chemical analogues (eg kinetin) are more widely used. As with auxins, applications of exogenous cytokinins to plants generally reduce secondary product yields (eg of *Datura* and *Papaver*). However, increases are also possible, for example caffeine from *Caffea* and sennoside from *Cassia augustifolia*).

GAs may influence secondary product yields

The main function of gibberellins (GA) is to regulate cell expansion. A large number of molecular species exist within this group and are based on the gibbane carbon skeleton. GA_3 is the most widely used example and it is now produced by fungal fermentation. Gibberellins were actually first discovered as metabolites of pathogenic fungi on rice. Of all plant growth regulators, gibberellic acids (GA) have been most widely reported to have effects on secondary product yield. For alkaloid-containing plants the extra internodal growth obtained is offset by the tissues having a lower metabolite content. GA applications can, however, have beneficial effects on glycoside production from *Digitalis* and rutin from buckwheat. With hops (*Humulus lupulus*) spraying the flowers with GA made them mature earlier and the flowers and fruits were more uniform, resulting in increased biomass yields. However, their α-resin content was reduced by 80% and the content of the volatile oil had been changed. Results with other crops vary, both increases and decreases being reported. At the moment therefore, we cannot predict the effects of gibberellins on secondary metabolite production and each system has to be tested.

Abscisic acid (ABA) is structurally related to the carotenoids and is usually regarded as a growth inhibitor. It has been shown to alter the distribution, but not to increase the overall yield of *Datura* alkaloids.

ethylene increases yield of latex

Ethylene (C_2H_4) is an unusual plant growth regulator in that it is a gas. It often stimulates the production of stress-related compounds (eg phytuberin) and increases the flow of rubber latex. It has now become common practice to apply ethephon to rubber trees to increase latex yields. Ethephon breaks down to release ethylene.

Π If the substances described above really regulate plant growth and metabolism, why is it not simple to apply them to increase secondary product yield?

There are a number of reasons why this might be so. First the concentration of the growth regulator may already be at an optimum level in the tissue. Further, it has been found that plant growth regulators interact with each other. A growth regulator may produce markedly different effects if applied in different combinations with other growth regulators.

empirical
approach is
often used

Finally, it has frequently been observed that even quite closely related strains of plant cells respond quite differently to particular growth regulator regimes. It must also be borne in mind that the application of exogenous plant growth regulators also affects the synthesis and turnover of endogenously produced growth regulators. Predicting the outcome of such application is, therefore, difficult and the approach used to successfully employ exogenously applied growth regulators is still largely an empirical one. In other words, we determine the effects of applying such growth regulators experimentally rather than following some universally applicable rules. Nevertheless, it is generally true that any enhancement of secondary product yield is usually an indirect one caused by an abnormally large growth of high producing tissues. The structures and activities of plant growth regulators are described in the BIOTOL text 'Crop Physiology'.

8.3.3 Bioreactor cultures

All our considerations so far have been of plants grown in soil. Special agricultural techniques may be used to increase yield but these are only modifications of the conditions to which wild plants are exposed. There is a complete alternative: the culture of cells or organs in large fermentation vessels or bioreactors. This approach is based on the widespread use of such methods for obtaining products from bacteria and fungi.

∏ Make a list of the potential advantages that would be obtained over conventional growing techniques by the adoption of such methods for culturing plants.

Here we cite three of the main advantages:

• greater availability of raw material. Some plants cannot, at present, be produced in sufficient quantities to satisfy demand because they are so difficult to cultivate. If their individual cells could be grown in large scale *in vitro* culture, then, in principle we would have limitless supplies of the raw material;

• lack of seasonality. Cell cultures can be grown at any time of the year without waiting for the appropriate planting season. This ensures a regular supply of the raw material;

• more consistent quantity and quality. Yearly fluctuations in weather etc cause variation in the secondary product content of plants cultivated in fields. Growth of the plant material in bioreactors means that they can be cultured in controlled environments. Thus yields would be more predictable and any extraction and purification procedures could be standardised. It would also be easier to maintain standards of purity of the product. This is especially important if the product is to be used for medicinal purposes.

∏ List some disadvantages of using bioreactors for the production of plant secondary metabolites.

You probably thought of the high cost (capital and running costs) of using bioreactors. Thus the cost of producing the raw materials may be much higher than field cultivation or collection of plants from the wild. Also, in many cases, the proportion of cell metabolism directed towards secondary metabolite production is much lower in cells cultivated *in vitro* than it is in cells grown in the wild (we will discuss this in more detail a little later).

In the 1950s and 60s studies of the mineral requirements of plants and the discovery of growth regulators led to the growth of small pieces of plants under axenic conditions (in the absence of competing micro-organisms). Since that time, a wide variety of procedures have been developed to cultivate plant cells and organs *in vitro* (see for example the BIOTOL text 'In Vitro Cultivation of Plant Cells'). We can broadly divide such cultures into two categories:

- the cultivation of differentiated cells and tissues (for example roots and shoots);

- the cultivation of undifferentiated cells (suspension cultures).

We will briefly examine each in turn.

Cultivation of differentiated tissue

shoot and root cultures

Our knowledge of the nutrient requirements of roots and shoots allows us to grow them in isolation of each other. Indeed the early work on the cultivation of excised roots provided the first evidence that the shoot was the site of synthesis of several vitamins. Shoot cultures are normally grown in medium made semi solid by the addition of agar, whereas most root cultures show much better growth in a liquid medium. In addition to the advantages mentioned above of the use of standard conditions, plant cultures offer the potential benefit of being easily able to control not only temperature, but also the exact constituents of the growth medium. Thus potential promoters of secondary product production can be easily added to the growth medium.

Whereas non-modified roots and shoots offer potential for the production of secondary products, there is great interest in the use of parts of plants that have been genetically transformed by species of the soil bacterium *Agrobacterium*. This bacterium is capable of transferring to plant cells plasmids which contain genes that upset the normal plant status of auxins and cytokinins. As a result either solid masses of cells (crown galls) or long tangled masses of roots (hairy roots) accumulate. The steps used in the production of hairy roots are shown in Figure 8.3.

∏ It might be helpful for you to read through the details given in Figure 8.3 and then to draw for yourself a flow diagram of the steps used to produce hairy roots. It will help you to remember this sequence.

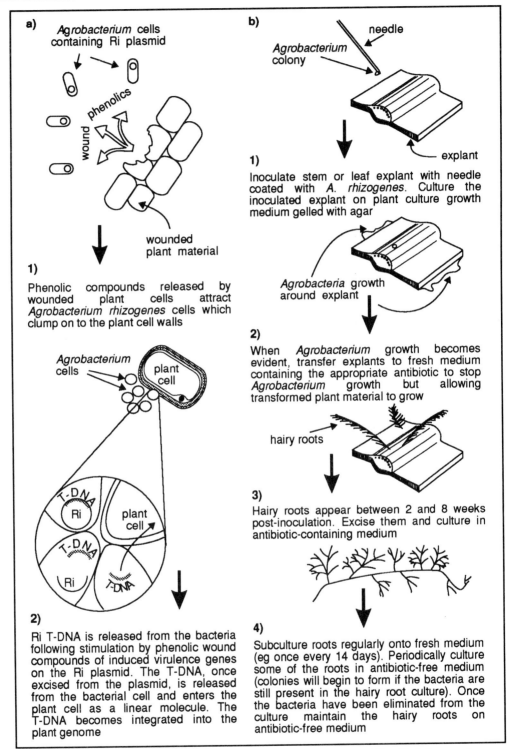

a)

Agrobacterium cells containing Ri plasmid

phenolics

wound

wounded plant material

1)

Phenolic compounds released by wounded plant cells attract *Agrobacterium rhizogenes* cells which clump on to the plant cell walls

Agrobacterium cells

plant cell

T-DNA

Ri

plant cell

T-DNA

Ri

T-DNA

2)

Ri T-DNA is released from the bacteria following stimulation by phenolic wound compounds of induced virulence genes on the Ri plasmid. The T-DNA, once excised from the plasmid, is released from the bacterial cell and enters the plant cell as a linear molecule. The T-DNA becomes integrated into the plant genome

b)

needle

Agrobacterium colony

explant

1)

Inoculate stem or leaf explant with needle coated with *A. rhizogenes*. Culture the inoculated explant on plant culture growth medium gelled with agar

Agrobacteria growth around explant

2)

When *Agrobacterium* growth becomes evident, transfer explants to fresh medium containing the appropriate antibiotic to stop *Agrobacterium* growth but allowing transformed plant material to grow

hairy roots

3)

Hairy roots appear between 2 and 8 weeks post-inoculation. Excise them and culture in antibiotic-containing medium

4)

Subculture roots regularly onto fresh medium (eg once every 14 days). Periodically culture some of the roots in antibiotic-free medium (colonies will begin to form if the bacteria are still present in the hairy root culture). Once the bacteria have been eliminated from the culture maintain the hairy roots on antibiotic-free medium

Figure 8.3 Transformation of plant material with *Agrobacterium rhizogenes*. a) Simplified diagram of events leading to the integration of T-DNA into the plant genome. b) Diagram of stages involved in generating hairy roots on a section of leaf lamina.

In the wild the occurence of crown galls or hairy roots constitutes a disease but biologists have exploited hairy roots as rapidly growing tissues which may be readily cultured for secondary metabolite production. They are especially suitable for substances normally made in the root tissue (eg *Datura* alkaloids). Special fermenter designs have been built for cultivating them (see Figure 8.4).

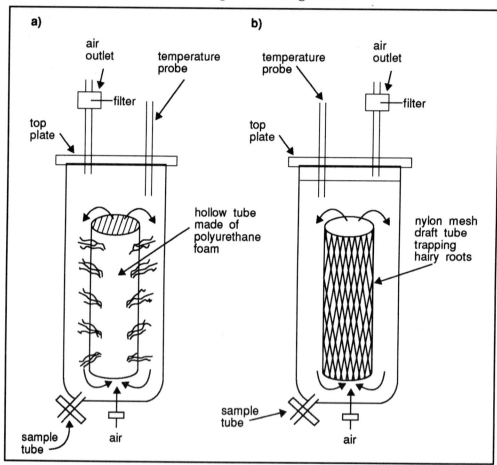

Figure 8.4 Airlift and bubble column bioreactors used to cultivate 'hairy root' cultures. a) In this bubble column a large column of polyurethane foam is used to anchor the roots; b) here the roots are trapped in a nylon mesh which forms the draft tube of the airlift bioreactor. Note that the device may be fitted with internal or external heating and cooling coils.

Cultivation of undifferentiated cells

problems arising from producing clumps of cells

Suspension cultures are composed of small clumps of undifferentiated (non-specialised) cells floating freely and proliferating in a nutrient medium. It is a characteristic of plant cells that they tend to stick together after cell division unless quite vigorously shaken or stirred. This can be a problem since large clumps are difficult to handle. Plant cells are many times bigger than fungal or bacterial cells. This, together with their rigid walls, means that they are readily torn apart by sheer forces caused by the stirring paddle blades of microbial fermentation vessels. Special designs must therefore be used. Stirring is also necessary to provide aeration, for the dissolved oxygen levels may be quite critical. Alkaloid production by *Berberis wilsoniae*, for example, reaches 10% of the cells' dry weight when the medium is 50% saturated with oxygen. This is reduced by two thirds when the level of oxygen reaches 60% saturation.

The growth rate of plant cells is relatively slow and months may be required to scale up the inoculum from a small flask to large vessels (up to 5000 litres). Secondary products are often not synthesised by actively dividing cells but by differentiated, or partially differentiated cells.

Π   See if you can think of a general way to overcome this latter problem, which is common to many cell cultures.

shikonin
production

The most obvious solution is to use a two phase process in which cells are allowed to grow in one medium and are then transferred to another medium in which secondary product formation takes place. This approach has been utilised for the manufacture of shikonin, an anti-inflammatory drug used widely in Japan (Figure 8.5).

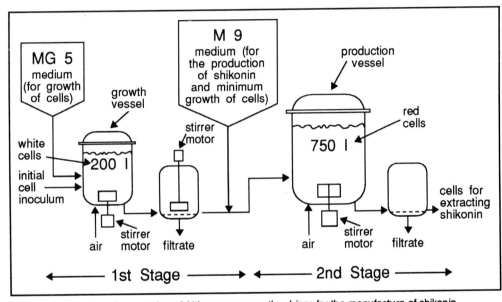

Figure 8.5 Two stage fermentation of *Lithospermum erythrorhizon* for the manufacture of shikonin.

Π   Bearing in mind that secondary product formation is low in rapidly growing and dividing cells but tends to increase when cells stop growing, see if you can suggest what the major differences are between the growth medium and production medium used in the fermentation of *Lithospermum erythrorhizon* shown in Figure 8.5. We do not expect you to give precise details of the differences.

You should have been able to argue as follows. The growth medium will require all of the ingredients needed for growth. In other words it must contain the full complement of mineral salts, a source of energy and carbon and the plant growth regulators needed to stimulate the growth of the cells (eg IAA or its equivalent). The production medium, on the other hand, does not need to be supplemented by growth regulators. At this stage, the major requirement is for a source of energy and carbon needed for the maintenance (not growth) of cells and for the synthesis of product.

Π   See if you can explain why the production vessel is larger than the growth vessel. What do the relative sizes of these vessels tell you about the residence time of cells in these vessels?

The aim of the shikonin process is to produce shikonin. Therefore in designing the process, the intention is to produce the cells that are required as quickly as possible. The faster the growth rates that can be achieved, the smaller the growth vessel that will be required. Once the cells have been produced, then attempts should be made to keep them healthy and productive for as long as possible. The sizes of the growth and production vessels shown in Figure 8.5 indicates that the cells are retained in the production vessel for about 3 times (actually 750 + 200) as long as they are in the growth vessel.

Shikonin is used in Japan:

- for treating wounds;
- as a red dye for silk;
- in cosmetics.

It is produced in the roots of *Lithospermum erythrorhizon* and cannot be chemically synthesised in an economic way. Its production in bioreactors is the most successful application of plant cell culture to metabolite production so far implemented. After 14 days of culture in production medium the cells are harvested and the shikonin is extracted. A single run of 750 l yields as much product as conventionally cultivated plants from a 176 400 m² plot (= 17.64 ha plot).

The M9 production medium contains only the normal mineral requirements required by the plants (Table 8.3). Thus the shikonin is produced by *de novo* synthesis.

Component	M9 medium (mg l⁻¹)	White's medium (mg l⁻¹)
Ca (NO₃)₂ 4H₂O	694	300
KNO₃	80	80
Na H₂ PO₄ 2H₂O	19	21
KCl	65	65
Mg SO₄ 7H₂O	750	750
NA₂ SO₄	1480	200
Mn SO₄ 4H₂O	0	5
Zn SO₄ 7H₂O	3	3
Fe₂ (SO₄)₃	0	2.5
Na Fe.EDTA	1.8	0
H₃ BO₃	4.5	1.5
K1	0	0.75
Cu SO₄ 5H₂O	0.3	0.001
Sucrose	30 000	20 000
Glycine	0	3
Nicotinic acid	0	0.5
Thiamine HCl	0	0.1
Pyridoxine HCl	0	0.1

Table 8.3 Composition (mg/l) of the M9 shikonin production medium and that of White's medium from which it was developed. M9 supports the production of about 12 times the shikonin produced in White's medium. IAA at 10⁻⁶ mol l⁻¹ was provided in both media and kinetin at 10⁻⁵ mol l⁻¹ in White's medium.

use of preformed precursors

An alternative strategy for producing secondary products *in vitro* is not to depend on the complete *de novo* synthesis of the product. Instead, the medium may be supplemented with preformed precursors of the product. A good example is the production of the spicy flavour compound capsaicin produced by *Capsicum frutescens*.

This compound is derived from valine and phenylalanine. Addition of these two amino acids to the medium increases the production of the desired agent.

use of
immobilised
and entrapped
cells

Plant cells may also be immobilised onto, or entrapped in, a solid matrix such as plastic foam or alginate gel for the production phase. This method works well if the product is secreted into the medium which is circulated through the mass of cells. The product may then be continually extracted and the cells maintained for an extended period.

The ability of immobilised cells to add compounds to the medium around them or remove them from it offers the possibility of using them as biological catalysts for biotransformation. Selected *Digitalis lanata* cells provided with β-methyldigitoxin convert it to β-methyldigoxin with good yields. This transformation involves site-specific addition of a hydroxyl group and the cells are used as a source of suitable enzyme. For such a technique to be useful the substrate must be available cheaply and reliably, the product must be valuable and the reaction must not be more easily performed chemically or by cultured micro-organisms.

∏ Do you think the use of plant cell cultures to produce secondary metabolites can be generally commercially viable?

Despite extensive research efforts, the products for which plant cell culture are actually used are very few in number. Many problems are encountered. Of particular importance are the slow growth rates, low yields and the decline in productivity as cultures age. Some plant species fail to produce their normal secondary products when grown in cell culture (eg *Cinchona ledgeriana*). However, in other cases metabolites not known in the intact plant may be obtained. The obstacles to the increased use of cultured cells are biological rather than technological. A greater understanding of what controls secondary metabolism is required in order to improve yields.

Before proceeding to the next topic, let us check your understanding of some of the above points with an SAQ and an intext activity.

SAQ 8.2

List at least three fundamental ways for the production of important chemical products.

∏ Examine Table 8.3 and list the major differences between the two formulations. From the information provided see if you can decide which differences may be the most important.

Relative to White's medium, M9 contains no kinetin (a cytokinin), three times as much calcium nitrate, seven times as much sodium sulphate, no manganese, three times as much boron, no iodide, glycine or vitamins, 30 times more copper sulphate and 1.5 times as much sucrose.

There is insufficient information to decide which of the differences is most important. You could have argued that the compounds omitted from White's had an inhibitory effect on shikonin production or that the additional amounts of other compounds could account for the greater production using M9. In practice, it has been shown that the major factors were the removal of inhibitory effects of glycine, thiamine hydrochloride and kinetin. Furthermore, the increase in copper concentrations in M9 relative to White's medium has also been shown to stimulate shikonin production.

8.4 Harvesting and purification techniques

∏ Is the time of harvesting crops critical in the production of secondary products?

You should have felt that the answer to this question is fairly obvious. You should have anticipated that the secondary product content of plants varies throughout their life cycle. For example, vanillin biosynthesis (in *Vanilla planifolia*) peaks eight months after pollination, and the morphine content of poppy capsules is highest 2½-3 weeks after flowering. In *Datura*, the hyoscine/hyoscyamine ratio falls from about 80% in young seedlings to 30% in mature flowering plants.

There may be seasonal fluctuations. Rhubarb contains only anthranols in winter, they are converted to anthraquinones when the weather warms up. Even the time of day of harvest can be important. This is often the case with cardiac glycosides or tropane alkaloids.

You should realise that many secondary compounds are not held at static concentrations but are converted from one form to another by the addition or removal of sugar residues. We may represent this dynamic state in the following way:

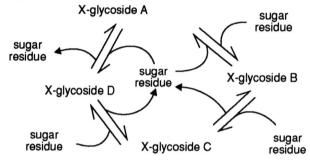

Thus, although the overall amount of the secondary product 'X' may remain constant, it is present in many different forms. The proportions of each of these forms may alter from hour to hour.

A wide range of harvesting techniques are employed, ranging from the picking of the leaves by hand to the use of sophisticated machinery. It may be necessary to separate flowers from leaves and stems. For underground organs, removal of soil is required. This may be achieved by washing or simply shaking if dry and sandy.

∏   See if you can suggest how the plant material should be treated after harvest.

reduction of water content

This very much depends upon the plant concerned. Large and fleshy roots are usually sliced to facilitate drying because all moist plant parts are liable to become spoilt by microbial attack. A reduction in moisture content also reduces shipment costs. A critical factor is the rate at which plant material is dried. The formation of desirable secondary products often depends upon enzymatic reactions taking place after harvest. For the development of fragrance from iris rhizomes (from *Iris spp.*) the rhizomes must be dried slowly. They are usually hung up on cords. Similarly, gentian roots are also dried slowly. They are heaped up after harvest and even covered with earth to allow secondary product formation. The characteristic flavour of vanilla pods develops only during curing (a slow drying process, the details of which vary around the globe).

In other cases, rapid drying is desirable to prevent enzymatic action. In tropical climates, drying in the heat of the sun may be quick enough although high humidity can be a problem. Artificial drying may be carried out using technology ranging from simple sheds containing fires to complex belt driers. The maximum temperature reached may be critical since too much heat can destroy the secondary metabolites. The

British Pharmacopoeia, for example, specifies the temperatures that should be used in certain cases (eg 65°C for *Colchicum* corms, 60°C for *Digitalis* leaves).

Π   Can you suggest any circumstances in which drying plant material after harvest would not be a good idea?

steam distillation

Contrary to the example of *Iris spp.*, there are numerous examples of fragrance production where drying is not always the appropriate first step. Volatile oils must often be extracted immediately after harvest to prevent loss. Steam distillation is the most common method. The distillate separates into oil and water layers. In some cases the oil saturated aqueous layer may be an article of commerce in its own right (eg rose water).

rectification efleurage

A second distillation step called rectification may be needed to remove resinous impurities (eg from the oil of caraway). In perfumery a process called efleurage is used, in which the oils are extracted from flowers into molten fats.

Plants that may be dried are usually subjected to a period in storage. Losses may also occur during storage.

Π   Try to make a list of what may cause the deterioration of stored products.

bio-deterioration

In some cases, eg cascara bark (from *Rhamnus spp.*), it 'improves' during years of storage. Generally, though, there is a gradual deterioration with time as the plants are subject to the attacks of fungi and bacteria, insects and rodents. Moulds are often the same as those responsible for the deterioration of food products (for example, *Mucor spp.* and *Penicillium spp.*). Of the insects, many species of beetles (particularly weevils) and moths are the main culprits causing the damage done to stored plant materials. Mites (Arachnida) are only microscopic but may be produced in vast numbers to cause considerable damage. Even in the absence of such organisms, stored plants change with time. There may be sufficient moisture within the plant tissue for enzyme action to continue, especially if the material is held in the warm. Direct sunlight can also encourage decomposition of plant chemicals.

Of course, the ideal situation would be to utilise crops immediately, but some storage time is almost inevitable and a great deal can be done by optimising the storage conditions. Dried plants gradually reabsorb moisture until they reach an equilibrium known as the 'air-dry' state. Digitalis, for example, must never be allowed to become air-dry if product yields are to be maintained. It has to be stored in the presence of a dehydrating agent. To prevent oxidation of sensitive metabolites, the air may be replaced with an inert gas. This is particularly applicable within bottles of extracted oils.

fumigation and hygiene

One method of discouraging microbial growth is to sterilise the plant material (for example by fumigation with ethylene oxide). Similarly, insecticidal and acaricidal (anti-mite) treatments may be used but this can cause problems of toxic residues. A better approach is to employ strict warehouse hygiene and possibly low temperature storage.

Π   Should it be of general concern that plants grown for their secondary products are subject to such deterioration in storage?

Let us consider the case when the product is to be used as a medicine.

The answer is 'yes'. Since drugs are to be consumed by sick people, strict quality control standards have been laid down and must be adhered to. These are concerned both with levels of impurities and with the content of the active molecules. Obviously unspecified detoriation of the plant material during storage may lead to the production of unspecified (and possibly undetected) impurities.

Following storage, use is finally made of the plants. Some dried plant tissues may be used directly. Many traditional medicines are used in this form. In many cases, however, purification of the secondary products is now required. The techniques employed depend on the characteristics of the molecules concerned. They may be extracted into water, alcohol etc and separated from impurities by partitioning against other solvents following changes of pH etc. Certain metabolites may then be changed by a series of chemical (or enzyme mediated) reactions into more useful substances. For example, steroidal hormones may be made in this manner from diosgenin extracted from yams. This partial synthesis is much easier than total synthesis, yet may provide more control over the composition of the final product than an entirely natural approach.

| SAQ 8.3 | Using the words provided below try to fill in the missing words in the following passage. |

The yield of secondary products can be greatly affected by the environment in which plants are grown. [   ], [   ], [   ] and [   ] may all be critical. Altitude is related to several of these factors. [   ] (grown for its insecticidal properties) is propagated on [   ] where it grows vigorously and then transplanted to [   ] where flowering is enhanced. Endogenous [   ] control plant development and sprays of the same chemicals can have profound effects on [   ]. Generally however, except in some cases, with [   ], secondary product yields are [   ] by their application. They may, however, be important in [   ] of [   ] or hairy roots. After harvesting, plants are often [   ]. The precise manner in which this is done may be critical in obtaining a satisfactory [   ]. Enzyme action is essential for the full development of [   ] fragrance but it must be [   ] as rapidly as possible in other cases. During [   ] plants grown for their secondary products are subject to losses caused by the attacks of [   ], [   ], [   ], [   ] and [   ].

Word List:

growth regulators; dried; orris; soil type; bioreactor culture; mites; storage; light intensity; rodents; prevented; fungi; Pyrethrum; day length; rain fall; bacteria; insects; plant morphology; suspended cells; lowland farms; high altitudes; giberellic acid; reduced.

8.5 Case studies

Here we will give a brief review of some specific secondary products

8.5.1 Ergot

Ergot is not a direct plant product, but is in fact the fruiting body of the fungus *Claviceps purpurea*. This is an obligate pathogen of a wide range of cereals and grasses but, because of the details of its anatomy, rye (*Secale cereale*) is particularly prone to infection. Ergotism or St Antony's fire is a disease caused by eating flour contaminated with the fungus. The natural life cycle of ergot is shown in Figure 8.6.

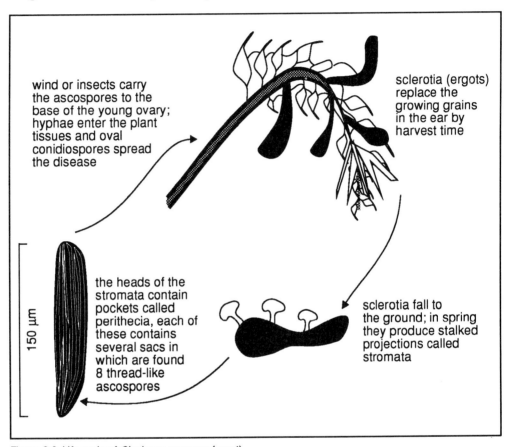

wind or insects carry the ascospores to the base of the young ovary; hyphae enter the plant tissues and oval conidiospores spread the disease

sclerotia (ergots) replace the growing grains in the ear by harvest time

the heads of the stromata contain pockets called perithecia, each of these contains several sacs in which are found 8 thread-like ascospores

150 μm

sclerotia fall to the ground; in spring they produce stalked projections called stromata

Figure 8.6 Life cycle of *Claviceps purpurea* (ergot).

use of sclerotia

This life cycle is exploited by Man commercially. Crops of rye are grown specifically for the purpose of producing ergot, mainly in central Europe, and a suspension of conidiospores is sprayed on at the optimal time. Sclerotia develop in many ears and the highest alkaloid content occurs just before normal rye harvesting time. The sclerotia are carefully dried and then stored at a cool temperature.

Whereas in earlier times crude extracts were used, nowadays only the purified alkaloids (mainly ergometrine, ergotoxine and ergotamine) are widely employed primarily in childbirth and to combat migraine. They may also be used as starting materials for semi-synthetic drugs. Note that alkaloids contribute only 0.5% of the weight of the dried sclerotia and attempts have been made to obtain the materials from fungal cultures.

The vegetative phase of the fungus can be cultured in large fermenters but sclerotia and their associated alkaloids are not produced by this method. However, a range of other useful substances are produced and, by the use of artificially selected strains, specific final products may be produced more efficiently than from wild strains. You should note that wild ergot produces materials which are of variable composition.

8.5.2 Pyrethrum

Pyrethrum is an extremely effective insecticide. It is produced from the dried flowers of *Chrysanthemum cinerariaefolium*, a perennial plant with an economic life of about three years. High yielding varieties are propagated by cuttings or micropropagation. Although indigenous to the Balkans it is now widely cultivated (particulary in Kenya and Ecuador). Altitude is important since a low night temperature stimulates bud production. 90% of insecticidal activity is present in the flowers so only these are collected, before the seeds have formed. This is possible for nine months of the year and is a labour intensive process. Plants are cut back at the end of the season to allow regeneration. The flowers are dried in hot air (before drying they are not toxic to insects). Pyrethrins are then either extracted and sold dissolved in a liquid or else the entire powdered flowers are used.

8.5.3 Opium

legal control of poppy cultivation

Opium is a product of the poppy *Papaver somniferum* which is also grown for its seed and its oil. Different cultivars may be used for each purpose. Cultivation of the poppies is usually under strict legal control. Poppies are best grown in warm temperate zones avoiding frost. Good rainfall is important early on but hot and dry weather is best once the plants are established. A rich soil is necessary. The seeds are very small and so must not be planted too deeply. The growing season lasts four or five months.

Traditionally, raw opium was extracted from the unripe seed capsules by cutting a small slit and collecting the coagulated latex the following day. Repeated several times, a single plant may yield 3-6g in this manner. This is clearly labour intensive so a more modern technique is now employed to extract the alkaloids from straw and dried seed capsules.

Raw opium has an illicit value as an addictive narcotic drug for eating or smoking. Careful storage is however required. It contains 25 alkaloids (including morphine, codeine, thebaine, noscopine, norceine and papaverine) together with organic acids, sugars, salts, proteins, pigments and water. The majority of legal opium is used for the extraction of its alkaloids but there are cases where its use *per se* is preferred since it exerts its effects more slowly. 90% of the morphine extracted is converted to related compounds such as codeine, ethylmorphine, pholcodeine. Because of problems of addiction, plants related to *Papaver somniferum* have been investigated as alternative sources of opiate type drugs.

8.5.4 Ginger

Ginger is important as a flavouring agent. It also has some medicinal properties (eg against motion sickness). It is produced from the rhizomes of *Zingiber officinale*, a reed-like plant grown in many subtropical parts of the world where the rainfall is greater than 80 mm per year. It can only be propagated vegetatively, pieces of rhizome being cultivated in a similar manner to potatoes. After the stems have withered the rhizomes are killed in boiling water and peeled, before being dried in the sun for 5-6 days. They are then moisturised and redried for a further period of 2 days. Ginger contains a huge array of secondary metabolites which contribute to its taste and aroma. Oil of ginger alone contains over 25 components. Much of the pungency is due to a range of phenolic compounds called gingerols.

gingerols

8.5.5 Liquorice

True liquorice consists of the dried roots and stolons of *Glycyrrhiza glabra*, a temperate leguminous plant which grows well in deep sandy, but fertile, soil near streams. It may be grown from seed or propagated vegetatively. After 3 or 4 years, the underground parts have grown sufficiently to be harvested. They may be sold peeled or unpeeled. The roots may be converted into black liquorice by a process of decoction (extraction by boiling) followed by evaporation. Different varieties are grown around the world and have markedly varied flavours. They are used in confectionery, tobacco and drug formulation.

decoction

8.5.6 *Catharanthus roseus*

Catharanthus roseus, the Madagascan periwinkle, is widely grown as an ornamental. Commercial supplies, for use as a drug, are obtained from both wild and cultivated populations. The plant has some local reputation for use against diabetes but, in contrast to many other species, its chief value has only been realised recently since its purified constituents have been investigated. *Catharanthus* contains a mixture of alkaloids which together have no therapeutic effect. In a highly refined form they are, however, effective against certain forms of cancer (eg Hodgkins disease and children's leukaemia). The alkaloids that are of particular interest are vinblastine and vincristine. These are dimers of two different alkaloids. They are present in much smaller amounts than many other secondary products that we have been considering. Vincristine constitutes only 0.0002% of the crude drug so large quantities of raw material and chromatographic purification are required. Much current research is aimed at increasing this yield using cell culture techniques, but so far the processes are not commercially-viable.

vinblastine and vincristine

Summary and objectives

In this chapter we have examined some of the factors which influence our selection of plant varieties and the strategies of cultivation we employ to cultivate crops for the production of secondary metabolites. We began by thinking about the use of the term yield before considering the techniques available to genetically manipulate strains to improve yield. We briefly examined the environmental factors which may influence yield. We also discussed alternatives to field cultivation for the production of plant metabolites. We considered the range of methods that may be employed to harvest and handle plant resources and examined the ways in which losses, both in terms of quality and quantity, may occur in stored plant materials. In the final part of the chapter, we briefly described some specific examples of important secondary plant products.

Now that you have completed this chapter you should be able to:

- list the different ways of expressing secondary product yield and appreciate how sometimes they can be misleading;

- suggest the reasons for selecting for cultivation one species rather than another;

- describe the various genetic techniques used for improving secondary product yield;

- list the environmental factors which affect secondary product yield and describe how these may be manipulated;

- explain why the large scale cultivation of plant cells *in vitro* for secondary product formation has currently found application only in a limited number of commercial operations;

- describe a range of methods used for the harvesting and subsequent handling of plants;

- explain how the losses in quantity and quality of product may occur during storage and suggest ways of minimising these.

Responses to SAQs

Responses to Chapter 1 SAQs

1.1   The answer is yes because there may be a non-uniformity which is less obvious than poor drainage.

1.2   There are various ways to do this. You could assign a number from 1-4 to each variety. Then, for each block, choose which variety goes into which plot by using a table of random numbers. However, for small numbers such as these it is easier to use a dice and ignore the numbers 5 and 6!

1.3   There is some scope for different designs in answer to this question. We have three levels of nitrogen fertiliser (20, 60 and 150 kg ha^{-1}) and two varieties, Norman (N) and Hobbit (H). The gradient in soil water content would suggest that a randomised block design would be most appropriate, stretching over the field towards the Eastern boundary. In each block there would be 6 plots to test each variety at each N level. Plots would be randomised within each block.

Below we have suggested two different arrangements based on the randomised blocks. Hopefully you used a design based on one of these. To randomise the arrangement of plots within these blocks, you could have used a dice. Since we have 6 experimental combinations (2 varieties x 3 N levels), we could have assigned each combination with a number.

For example, H(20) = 1, H(60) = 2, H(150) = 3, N(20) = 4, N(60) = 5 and N(150) = 6

(where H = Hobbit, N = Norman and numbers in brackets refer to levels of application of the nitrogen fertiliser.

Then by rolling the dice, we could decide which combination to place in each plot.

Design 1:

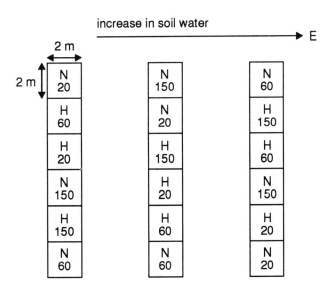

Design 2:

increase in soil water

2 m →→→→→→→→→→→→→→ E

2 m	N 20	H 60	H 20		N 150	N 20	H 150		N 60	H 150	H 60
	N 150	H 150	N 60		H 20	H 60	N 60		N 150	H 20	N 20

1.4   Calculate RGR using Equation 1.2 and AGR is calculated from $W_2 - W_1/t_2 - t_1$.

Quantity	Plant 1	Plant 2	Units
W_1	2	20	g
W_2	5	30	g
t_1	14	14	days
t_2	21	21	days
$W_2 - W_1$	3	10	g
$\ln W_2 - \ln W_1$	0.916	0.405	
$t_2 - t_1$	7	7	days
AGR	0.43	1.42	g day^{-1}
RGR	0.13	0.057	g g^{-1} day^{-1}

Note that plant 2 has grown the most over the 7 day period (AGR = 1.42 g day^{-1}) but is growing more slowly when related to its initial weight (RGR = 0.057 g g^{-1} day^{-1} compared to 0.13 g g^{-1} day^{-1} for Plant 1).

1.5   1) ULR = 8.5×10^{3} g cm^{-2} day^{-1}.

This can be calculated from:

$$ULR = \frac{(W_2 - W_1)(\ln A_2 - \ln A_1)}{(A_2 - A_1)(t_2 - t_1)} \qquad \text{(see Equation 1.4)}$$

2) LAR on day 35 = 106 cm^2 g^{-1}, on day 42 = 119.7 cm^2 g^{-1}.

These are calculated from:

$$LAR = \frac{A}{W} \qquad \text{(see Equation 1.5)}$$

(day 35 = $\frac{106}{1.0}$ and day 42 = $\frac{467}{3.9}$).

3) The mean (average) LAR = 114 cm^2 g^{-1}.

This can be calculated from the equation:

$$LAR = \frac{(A_2 - A_1)\,(\ln W_2 - \ln W_1)}{(W_2 - W_1)\,(\ln A_2 - \ln A_1)} \qquad \text{(see Equation 1.6)}$$

$$= \frac{(467-106)(\ln 3.9 - \ln 1.0)}{(3.9-1.0)(\ln 467 - \ln 106)}$$

1.6   CGR = 2.92 g m^{-2} day^{-1},

LAI (1st harvest) = 1.3,

LAI (2nd harvest) = 2.2.

To calculate CGR we use Equation 1.9:

$$CGR = \frac{W_2 - W_1}{t_2 - t_1}$$

Remember, however, that the values for W have to be expressed on the same basis (ie the same area).

In this case W$_1$ = 75 g per 1 m^2 and W$_2$ = 200 g per 1.5 m^2.

Thus W$_2$ = 133.3 g per 1 m^2.

Substituting in:

$$CGR = \frac{133.3 - 75}{60 - 40} = 2.92 \text{ g m}^{-2} \text{ day}^{-1}.$$

To calculate LAI we use the relationship:

$$LAI = \frac{A}{P} \qquad \text{(see Equation 1.8)}$$

Thus at harvest 1:

$$LAI = \frac{1.3 \text{ m}^2}{1 \text{ m}^2} = 1.3$$

(notice the units cancel);

and at harvest 2:

$$LAI = \frac{3.3 \text{ m}^2}{1.5 \text{ m}^2} = 2.2.$$

Responses to Chapter 2 SAQs

2.1

1) Stomatal resistance (r_s). This can vary from 50 - 10 000 s m^{-1}.

2) Under conditions of water limitation. Under such conditions the stomata will be closed and resistance will be at its highest.

3) Mesophyll resistance (r_m). When the stomata are fully open the stomatal resistance will be approximately 50 s m^{-1}. Thus the resistances can be ranked in the order:

$r_m \sim 100\text{-}1000$ s m^{-1}

$r_s \sim 50$ s m^{-1}

$r_b \sim 20$ s m^{-1}

The greatest resistance to CO_2 diffusion under the conditions specified is, therefore, mesophyll resistance (r_m).

2.2

The line you should have added should be similar to that shown below.

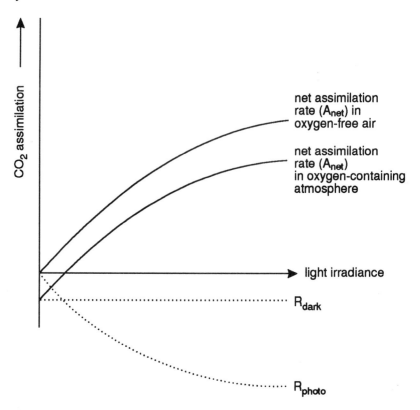

In an oxygen free atmosphere, the rate of respiration (both dark respiration and photorespiration) would be reduced to virtually zero. Therefore, there would be little dissimilation of CO_2 (ie R_{dark} and $R_{photo} \sim O$) and the net assimilation rate would become the gross assimilation rate. In practice R_{dark} and R_{photo} would only be reduced to zero for a very short period, the photosynthetic generation of O_2 would quickly enable both dark and photorespiration to become established despite the atmosphere being initially free of oxygen.

2.3 At low light irradiances, photorespiration in C_3 leaves would be very low. We might anticipate therefore that, under these conditions, C_3 and C_4 leaves would behave very similarly to each other. However, remember that C_4 plants consume energy to transport CO_2 to the site of photosynthesis (see Figure 2.6). Thus there is a greater demand for respiratory energy in C_4 plants and this lowers the net CO_2 assimilation rate.

2.4 The data illustrated in Figure 2.12 indicate that C_3 crops will benefit from increasing the CO_2 concentration over the range 300-600 ppm but C_4 plants will show very little, if any, increase.

2.5 1) Increased. The increased permeability of the membranes would mean that the concentration gradients would be more easily dissipated by diffusion. More energy would be consumed to maintain these gradients.

2) Increased. The virus is multiplied within the tissue and energy would be consumed to do this. Some energy is also used to repair the damage caused by the virus and on the defence mechanisms. The consumption of this energy would not be coupled to increased yields of the plant biomass.

2.6 Leaves are metabolically more active than the other organs. They contain many more metabolically active enzymes which are constantly being turned over. Furthermore the exposure of leaves to UV radiation causes damage to DNA that costs energy to repair.

2.7 1) 10 kg day^{-1}.

Energy consumed = weight of crop x maintenance energy coefficient (α). Using the value for α given in Table 2.4:

$= 1000 \times 0.01 \text{ kg day}^{-1}$

$= 10 \text{ kg day}^{-1}$.

2) 60 kg day^{-1}.

The maintenance energy coefficient for oil rich seed crops is $0.03 \text{ kg}^{-1} \text{ day}^{-1}$ (see Table 2.4) at $20°C$.

At $30°C$, this will be approximately $2 \times 0.03 \text{ kg kg}^{-1} \text{ day}^{-1}$.

Therefore $\alpha_{oilseed}^{30°} = 0.06 \text{ kg kg}^{-1} \text{ day}^{-1}$.

The consumption of carbohydrate for maintenance purposes will, therefore, be:

$= 1000 \times 0.06 \text{ kg day}^{-1}$

$= 60 \text{ kg day}^{-1}$.

3) For example 1 (potatoes) 0.01 kg kg^{-1} are consumed per day

$$= \frac{0.01}{1} \times 100\% = 1\% \text{ day}^{-1}.$$

For example 2 (oilseed rape) 0.03 kg kg^{-1} are consumed per day at 20°C,

= approximately 0.06 kg kg^{-1} day at 30°C.

Thus the proportion consumed is $\frac{0.06}{1} \times 100\% = 6\% \text{ day}^{-1.}$

Note the data was taken from Table 2.4.

These calculations should have convinced you that maintenance energy requirements have a significant impact on the net biomass yield of a crop.

2.8 In O$_2$-free air, respiration would be very low and the plant would be expected to suffer from the lack of maintenance respiration.

2.9 The order you should have written is:

oil rich seed crops	high in fats, therefore low CVF
protein rich seed crops	high in protein, intermediate CVF
cereals	high in carbohydrate, some protein, therefore fairly high CVF
root/tuber	high in carbohydrate, low in other components, therefore high CVF

Some typical CVFs for these crop types are:

oil rich seed crops; 0.50 g g^{-1};

protein rich seed crops; 0.65 g g^{-1};

cereals; 0.70 g g^{-1};

root/tuber crops; 0.75 g g^{-1}.

2.10 A1 with B2.

A2 with B4.

A3 with B3.

A4 with B1.

Responses to Chapters 3 SAQs

3.1

1) Climate.

2) Weather.

3) Weather. (You might have considered this to be climate, especially if these amounts of rainfall in June are a regular occurrence).

4) Climate.

3.2
There are quite a lot of issues you could have commented upon. Here we will include the major features.

With the lower LAD value, this means that there is a lower leaf area per unit volume of canopy. Thus, light penetrates further into the canopy (see Q_n in Figure 3.3). Also, less CO_2 is absorbed and thus there is a slightly higher CO_2 concentration in the less dense canopy. Also, there is less transpiration and thus the humidity in the canopy is lower. This effect is also added to by the free movement of air in the less dense canopy (see wind line in Figure 3.3). With a dense canopy the lack of air movement also leads to a greater retention of heat within the canopy. Thus, generally the temperature in a dense canopy is higher than in a less dense one.

3.3
$E = 181$ kJ mol^{-1}

since $E = \dfrac{hc}{\lambda}$ = (6.63 x 10^{-34}) (3 x 10^{8})/6.6 x 10^{-7} = 3.01 x 10^{-19} J photon^{-1}.

Therefore E mol^{-1} = (3.01 x 10^{-19} J photon^{-1}) x (6.02 x 10^{23}) = 181 kJ mol^{-1}.

If you forget to convert the wavelength from nm to m your answer will be out by 10^{9}.

3.4
Your graph should have looked like this:

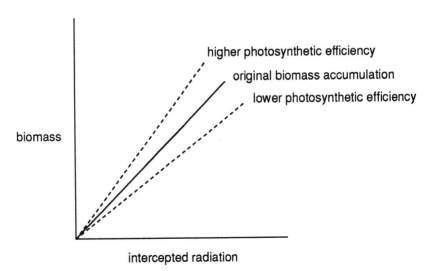

3.5   Let us consider the LAI = 1 curve first. With a single leaf layer, we would anticipate that the gross CO_2 assimilation would increase as the incident PAR increases. However, at 200-250 J m^{-2} s^{-1} PAR, these leaves will be light saturated. Any further increase in light intensity will not lead to a further increase in CO_2 assimilation rate.

Now let us consider the LAI = 5 curve. We would anticipate that the relationship between gross CO_2 assimilation rates and incident PAR of the upper leaf layer to be similar to that of an LAI = 1 canopy.

However as light intensity increases, more and more of the radiation will be transmitted to lower layers of the canopy. This will lead to increased CO_2 assimilation by the lower layers. Thus the overall CO_2 assimilation rates continue to rise even when the incident PAR is greater than 200 J m^{-2} s^{-1}. In theory we might anticipate that the maximum gross CO_2 assimilation rate of a LAI = 5 canopy would be 5 times that of a LAI = 1 canopy. In practice, this is not achievable since only a small portion of the incident radiation is transmitted to lower leaf canopy layers. The actual amount of light transmitted to these lower layers is greatly influenced by the morphology of the leaves and of their physical arrangements. These, of course, influence the shapes of the curves shown in the figure associated with this question.

3.6   Since a high N-content favours shoot growth, it makes sense to add N-fertilisers to root crops early. This will help the plants to quickly establish the 'photo-harvesting' canopy which will lead to improved CO_2 assimilation. Late addition of N-fertilisers should be avoided as this will divert photosynthates to shoot rather than root production.

3.7   The data indicate that when the prices of fertilisers are low and the prices of crops are high, then there is an increased use of fertilisers. This increased use of fertilisers is indicated by the higher yields at higher price ratios shown in Figure 3.21.

3.8   Water use efficiency (WUE) = transpiration rate/photosynthetic rate.

WUE = 50/0.6 = 83.3 mg H_2O mg^{-1} CO_2 = 83.3 g H_2O g^{-1} CO_2 fixed.

3.9   The answer is probably yes. If the soil is held at its field capacity, it means that water is readily available to plants growing in the soil. Thus we would anticipate that the field would, provided suitable nutrients were available, readily support plant growth.

Responses to Chapter 4 SAQs

4.1   Species A appears to be a saprophyte because it was found only on dead plant material. Since it is only found on detached leaves, it appears unlikely that species A is actually responsible for killing the plant material.

Species B could be either a necrotroph or a saprophyte. The presence of the fruiting bodies on dead parts of plants that are otherwise alive may indicate that the species is capable of killing the plant tissues and then growing on the dead tissues. Thus it is tempting to suggest that this is a necrotroph. However, further study would be required to rule out the possibility that it was a saprophytic fungus.

Species C could be a biotroph since it appears to be growing in living host tissues.

4.2   There are several opportunities to control *Puccinia graminis*. First we could deal with cultural control. By removing barberry plants from the vicinity of wheat fields, the pathogen is unable to complete its life-cycle, thus the incidence of the disease should be dramatically reduced.

Similarly removal and destruction of wheat straw at the end of the season will remove the overwintering teliospores. Thus, in spring, few basidiospores will be produced and this in turn would lead to few aecidiospores and urediospores.

As an alternative to these cultural control methods, wheat varieties which are resistant to the black rust disease-causing organism might be cultivated.

Finally we may consider chemical control methods. Obviously we need to use a fungicide which is effective against *P. graminis* but not detrimental to the host plant. The resting teliospores are in the overwintering state, resistant to most chemical sprays. It is the rapidly growing stages of spring and summer which are more susceptible.

It is not usual to attempt to kill the disease organism on barberry using fungicides. This is a relatively short period of the life cycle of the disease-causing organisms and the barberry plants, being wild are often well scattered and not easily accessible. Thus it is usual to spray the wheat plants directly. But when should this be done?

If the fungus is killed early in the season then the damage it will cause will be minimised. We must not, however, spray the wheat too early as the wheat may become re-infected with late transfers of aeciospores from the spring hosts. Obviously the maturation of the aeciospores will be dependent upon the climate and may be different from year-to-year. Thus judgement as to the best time to spray is a skilled activity and must be made using knowledge of the local conditions.

4.3   There are fewer opportunities to control late blight disease (*Septonia apiicola*) in celery than there are with black rust diseases. There are, for example, no alternative hosts for *S. apiicola*, so this disease cannot be controlled by removing such hosts. In this case the disease is best controlled by using a combination of strategies.

First, any diseased plants should be removed and destroyed. If possible a crop rotation should be employed to reduce the probability of the disease being carried over from one year to the next. Chemical control agents could be used to spray crops which begin to show signs of the disease. If available, resistant varieties should be used. Avoid the use of tools which have been used in infected areas or decontaminate such tools before using them.

In practice there may be many limitations associated with each of these measures. For example, removing diseased plants may be labour (and therefore cost) intensive. Chemical control agents that are used must be safe to the consumer. Their use also incurs an economic cost. Resistant varieties may show poor growth characteristics.

With celery, in contrast to cereals, the whole plant is harvested. The roots and leaves are trimmed from the harvested plants prior to marketing. It is important that the discarded plant parts are removed from the site of cultivation.

4.4   From the data presented in the question, it appears that the grain mite will not grow below 5°C nor above 35°C and does not survive if the relative humidity is less than 62%. Its tolerance to relative dryness is, however, influenced by temperature.

It would appear that we have these main options to reduce grain mite damage to stored grain. We can:

- keep the grain below 5°C (ie refrigerated);

- keep the grain above 35°C;

- keep the grain dry.

In practice it would be extremely expensive to keep the grain in a refrigerator or at temperatures above 35°C. In practice, of course, a combination of processes might be used. For example, warming the grain soon after it has been harvested to, for example, 35°C will kill any *Acarus* present and aid drying. If this grain remained dry, then it should remain free of the mite providing mites are prevented from entering the storage containers.

4.5   Number of grains produced per unit area = ears ha^{-1} x grain/ear.

Yield per ha = grain ha^{-1} x mean mass of grain (g ha^{-1}).

= grains ha^{-1} x mean mass of grains + 10^6 (t ha^{-1}).

Thus:

Plant Density (10^6 ha^{-1})	1.99	2.74	3.29	4.35	5.21
No. grains ha^{-1} (x 10^{-6})	140	144	149	153	157
Yield of grains (t ha^{-1})	5.17	5.13	5.46	5.47	5.33

Graphically:

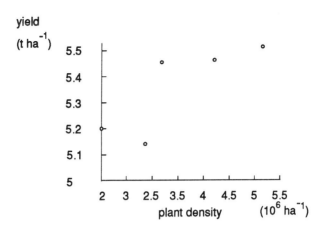

You will immediately recognise that an over 2½ fold increase in plant density is not matched by a similar increase in yield. In fact within the likely experimental error associated with such analysis, there is barely a 10% increase in yield. Thus we can see that in the range 2-5.5 x 10^6 plants ha^{-1}, there is virtually no increase in yield.

Inspection of the other data shows that the number of grains per ear are virtually unchanged at different plant densities (21.7-23.8/ear). Furthermore, the weights of individual grains are also largely independent of plant density (range 35.2 to 36.9 g per 1000 grain).

The factor which seems to be most effected by plant density is the number of ears per plant.

Graphically:

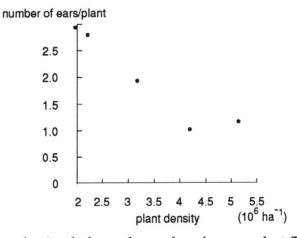

The higher the plant density, the lower the number of ears per plant. Thus the increase in plant density is more-or-less counteracted by the reduction of the number of ears per plant. The result is that the number of ears ha^{-1} is not substantially different for any of the plant densities analysed.

We must conclude, therefore, that increasing the plant density above about 2.0 x 10^6 plants ha^{-1}, leads to very little increase in yield because the number of ears per plant decreases.

You could have used your expertise in statistics to tackle this SAQ.

Responses to Chapter 5 SAQs

5.1   Radiant heat will penetrate the glasshouse and will be absorbed by the sensor causing errors in the estimation of air temperature. The sensor should be shielded from direct radiation with a protective housing designed to allow free air circulation. The positioning of the temperature sensors is also important bearing in mind the objective which is to ensure a uniform and fairly constant temperature within the plant growing zone. A number of detectors, depending on the size of the facility, should be distributed in this zone.

5.2   Plants close to the window would be exposed to much greater temperature variation, far higher temperatures when the heating is on and lower temperatures when the heating is off, than those in the centre. Circulation fans would be an obvious addition to the equipment. In the absence of fans the problem is tackled on a small scale by moving the plants periodically so that they each have a period close to the glass. This, of course, is impractical on a large scale.

5.3   The loss of heat from the gables for a width of 11 m = 7618

From the walls (1.83 m as the curtain wall is 0.61 m) =

2637 (6.1 m) + 7032 (15.2 m) + 4993 (11.0 m) = 29 324 (this is x 2 to account for both sides at 21.3 m and both ends)

From the roof = 21.3 x 3047.2 = 64 905.36

Total = 7618 + 29 324 + 64 905.36 = 101 847.36

Correct for climate (K), structure (S) and material (C) =

101 847.36 x 0.88 (K) x 0.70 (S) x 1.0 (C) = 62 738

Curtain wall (0.61 m) = 879 (6.1 m) + 2334 (15.2 m) + 1664 (11.0 m) = 9774 (again doubled to account for both sides and ends.

Correct for climate (K) and material (C) =

9774 x 0.88 (K) x 0.46 (C) = 3956.5

Therefore the combined heat loss is:

62 378 + 3956.5 = 66 334.5 W hr^{-1}

5.4   Equation 7.2 tells you that the amount of heat lost depends on the density of air, so as the elevation increases the density of air decreases and therefore the cooling capacity of a given volume also decreases. Consequently, greater volumes of air are required to achieve the same cooling effect. An increase in light irradiance will also increase the heat input, again requiring an increase in the rate of air flow.

5.5 Light generates heat which might exceed the capacity of the refrigeration unit. This is avoided by providing a separate compartment for the lights, which can be cooled against the ambient temperature by the use of a fan. The optimum temperature for lamp operation is also much higher than that tolerated by plants and the separation of growth and light compartments enables the lights to be operated at maximum efficiency.

5.6 Low pressure sodium (LPS) lamps emit almost exclusively in the 560-610 nm band. It is to be expected, therefore, that growth will be similar to, but more extreme than, fluorescent and HPS lamps which are also deficient in wavelengths from 680-780 nm. Extra deep green foliage, larger and thicker than with other sources, and very thick stems develop; elongation is slowed, multiple side shoots develop even on secondary shoots and flowering occurs but stalks do not elongate.

5.7 A 2% reduction in the light reaching the crop can be directly translated into a 2% loss of income, therefore 2% of £25 is 50p. In order to recoup this from the fuel costs, a saving of greater than 50p in £4 or 12.5% is needed. This may seem a facile calculation but it is included to illustrate the point that very large savings are required to make energy saving measures cost effective. A situation that is unlikely to change unless fuel costs rise dramatically.

5.8 The respiration rate could increase such that the LCP is raised to a level that is not met by the low irradiance. As a consequence there will be a net loss from the crop. This can be avoided by keeping the plants at a lower temperature, or using supplemental lighting when the daylight intensity falls below a critical level.

5.9 The tomato mosaic virus is capable of adapting rapidly to overcome the resistance of the plant. There were five strains of TMV identified by the mid 1970s. Strains of tomato have been developed which have varying degrees of resistance. The incidence of TMV would, therefore, be expected to rise and control of infection cannot be guaranteed unless new strains of tomato were developed which were resistant to the newly adapted strains of TMV.

Responses to Chapter 6 SAQs

6.1 Clear plastic will allow light to penetrate and this will encourage the growth of algae, which would utilise some of the nutrients as well as possibly release toxic compounds. This is minimised by using black plastic. The use of plastic which is white outside allows the upward reflection of light which is beneficial for the growth of the crop.

6.2 One benefit of the NFT system is the good aeration that occurs by virtue of the thin film of nutrient. Stationary pools of nutrient will become depleted of oxygen which is detrimental to roots.

Stationary pools will also interfere with the supply of nutrients so the system will not be being used at its best.

6.3 Such a design might achieve the doubling of yield but it is difficult to see how the lettuces could be harvested. One consideration is to begin to harvest at one side of the greenhouse and to dismantle the frame as you go. However each A frame has a nutrient reservoir beneath it. Unless this was covered in some way, the operator would have nothing to stand on while doing the harvesting. That having been said, the nutrient reservoir need be no more than a trough, gently sloping to the centre, so modified designs might be feasible.

6.4 Method a) will lead to some root damage irrespective of how careful the operator is, whereas root damage should be minimal using method b). Root damage at pricking out often results in a check to growth.

6.5 Chlorine is the missing element. The minerals will be dissolved in tap water which contains more than an adequate supply of chlorine.

6.6 A basic procedure would be to sample always at the same time ie just before or just after the operation of the drip irrigation, and from a predetermined position ie between or beneath plants.

6.7 Remember that rockwool slabs are not sealed in plastic and excess fluid can simply drain out of them. The reduction in conductivity depends on the volume of drip irrigation applied. The conductivity can be reduced by either flushing for a longer period or by reducing the conductivity of the irrigation solution.

6.8 The higher conductivity of the solution would reduce its water potential, and this is likely to reduce the growth of the plant. This is, in fact, what happens.

6.9 1) Applies, because hydroponic systems can be built anywhere.

2) applies; because of the greater control, water and minerals are used more efficiently.

3) Does not apply; question of cost is a complex one because of the large number of inputs but hydroponics *per se* is more expensive than OFA.

4) Does not apply specifically to hydroponics

5) This applies; crops are usually grown at higher density than in OFA and so less land is used for the same number of plants.

6) This applies; there is only limited control of this in OFA.

6.10   Developing fruits are stronger sinks than roots for the products of photosynthesis. Thus roots might actually be, at least partially, starved at this stage. It is conceivable that this reduces root vigour and increases their susceptibility to attack.

Responses to Chapter 7 SAQs

7.1

1) **Primary metabolites**

- glucose, important monosaccharide;
- chlorophyll, key photosynthetic pigment;
- hexokinase, a major respiratory enzyme.

Secondary metabolites

- pyrethrin, an ester with insecticidal properties;
- digitoxin, a drug (an alkaloid) used for heart complaints;
- vincristine, a drug with antitumour properties;
- atropine, a drug with effects on the nervous system.

Lead is not a metabolite and was our trick item.

Do not worry if you did not recognise all of the specific examples of secondary products. We shall meet them again later in the chapter.

2) Carotenoids are involved in photosynthesis and could be considered as a primary metabolite. But, in some instances they are also important colouring agents and may be considered to be secondary metabolites. The point we are trying to make is that it is sometimes difficult to distinguish between primary and secondary metabolites.

7.2

Drug name	Plant of origin	Effect	Type of drug
DAP 30	Carnation (*Dianthus caryophyllus*)	antiviral, antitumour	protein
morphine	Opium poppy (*Papaver somniferum*)	relieves severe pain, hypnotic, sedative, decreases metabolism	alkaloid
taxol	Pacific yew (*Taxus brevifolia*)	antitumour	alkaloid
tubocuarine	Cuare (*Strychnos* sp., *Chodrodendron* sp.)	neuromuscular relaxant	alkaloid
digitalis	Foxglove (*Digitalis* sp.)	strengthens heart beat, promotes diuresis	cardenolide glycoside
atropine	Thornapple (*Datura spp.*), Heribane (*Hyoscyamus spp.*), Deadly nightshade (*Atropa belladonna*)	antagonises acetylcholine, stimulates CNS, dilates pupils, reduces bronchial secretions	alkaloid
quinine	Cinchona (*Cinchona spp.*)	antimalarial, bitter tonic and stomachic	alkaloid
ergotamine	*Claviceps purpurea* (a fungal parasite of cereals)	vasoconstrictor, stimulates uterine muscles	alkaloid

NB: Digitalis is a general terms for alkaloid preparations from the foxglove. Names such as digitoxin refer to a specific combination of cardenolides and sugars.

7.3

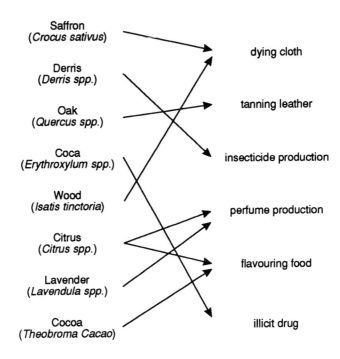

Responses to Chapter 8 SAQs

8.1

fresh weight per m² per month	$\frac{2.4}{3} = 0.8 \text{ kg m}^{-2}$
dry weight per m² (one harvest)	10% of 2.4 = 0.24 kg = 240 g m⁻²
dry leaf weight per m² (one harvest)	$\frac{240}{2} = 120 \text{ g m}^{-2}$
alkaloid produced per m² (one harvest)	1% of 120 = 1.2 g m⁻²
alkaloid produced per total dry weight	$\frac{1.2}{240} = 0.005 \text{ g g}^{-1}$ dry wt
alkaloid produced per total fresh weight	$\frac{1.2}{2400} = 0.005 \text{ g g}^{-1}$ fresh wt
alkaloid produced per m² per month	$\frac{1.2}{3} = 0.4 \text{ g m}^{-2}$

8.2   The list we hoped you produced is:

a) for the production of chemicals by *de novo* synthesis;

b) for the production of chemicals incorporating pre-formed organic precursors;

c) for the production of specific chemicals using cells as biological catalysts to add (or remove) substituents to a specific molecule in a specific place and/or configuration.

You may have included the use of immobilised cells in your list. Technically all of the systems listed above could employ the use of immobilised cells, then the use of immobilised cells is not considered to be one of the three fundamental production methods.

8.3   The yield of secondary products can be greatly affected by the environment in which plants are grown. [Soil type], [day length], [light intensity] and [rainfall] may all be critical. Altitude is related to several of these factors. [Pyrethrum] (grown for its insecticidal properties) is propagated on [lowland farms] where it grows vigorously and then transplanted to [high altitudes] where flowering is enhanced. Endogenous [growth regulators] control plant development and sprays of the same chemicals can have profound effects on [plant morphology]. Generally however, except in some cases, with [gibberellic acid], secondary product yields are [reduced] by their application. They may, however, be important in [bioreactor culture] of [suspended cells] or hairy roots. After harvesting, plants are often [dried]. The precise manner in which this is done may be critical in obtaining a satisfactory [product]. Enzyme action is essential for the full development of [orris] fragrance but it must be [prevented] as rapidly as possible in other cases. During [storage] plants grown for their secondary products are subject to losses caused by the attacks of [fungi], [bacteria], [insects], [mites] and [rodents].

Suggestions for further reading

Related BIOTOL texts

Crop Physiology, Butterworth-Heinemann, Oxford, ISBN 0 7506 0560 X

Techniques for Engineering Genes, Butterworth-Heinemann, Oxford,
ISBN 0 7506 0556 1

Strategies for Engineering Organisms, Butterworth-Heinemann, Oxford,
ISBN 0 7506 0559 6

In vitro Cultivation of Plant Cells, Butterworth-Heinemann, Oxford, ISBN 0 7506 0554 5

Biotechnological Innovations in Crop Improvement, Butterworth-Heinemann, Oxford,
ISBN 0 7506 1512 5

Defence Mechanisms, Butterworth-Heinemann, Oxford, ISBN 0 7506 0565 0

Statistics

Fowler, J.A. and Cohen, L., Practical Statistics for Yield Biologists, Wiley, Chichester,
ISBN 0 3350 9207 1

Parker, R.E., Introductory Statistics for Biology, Cambridge University Press,
Cambridge, ISBN 0 5214 2718 9

Sokal, P.R. and Rohlf, F.J., Biometry, 2nd Edition, Freeman, New York,
ISBN 0 7167 7254 7

Watt, T.A., Introductory Statistics for Biology Students, Chapman Hall, London,
ISBN 0 4124 7150 7

Underpinning physiology/biochemistry

Dale, J., Plants and Water, Cambridge University Press, Cambridge,
ISBN 0 5214 6993 7

Edwards, G. and Waller, D., C_3 C_4 Mechanisms, Cellular and Environmental
Regulations of Photosynthesis, Blackwell, Oxford, ISBN 0 6320 07672

Hall, D.O. and Rao, K.K., Photosynthesis, Cambridge University Press, Cambridge,
ISBN 0 5214 2806 8

Lewis, O.A.M., Plants and Nitrogen, Cambridge University Press, Cambridge, ISBN 0 5214 2776 2

Shrewry, P.R. and Stobert, K., Seed Storage Compounds, Oxford University Press, Oxford, ISBN 0 4240 0131 4

Taiz, L. and Zeiger, E., Plant Physiology, Addison and Wesley, New York, ISBN 0 8053 0245 A

Crop production and controlled environments

Downs, R.J., Controlled Environments for Plant Research, Colombia University Press, New York

Duke, J.A., CRC Handbook of Alternative Cash Crops, CRC Press, New York, ISBN 0 8493 3626 1

Enoch, H.Z., Carbon Dioxide Enrichment of Greenhouse Crops, CRC Press, New York

Mastalerz, J.W., Greenhouse Environment, Wiley, Chichester

Nelson, P.V., Greenhouse Operation and Management, 3rd Edition, Reston Publishing Co, London

Pratley, J.E., Principles of Field Crop Production, Oxford University Press, Oxford, ISBN 0 4240 0131 4

Rees, A.R., Cockshull, K.E., Hand R.G. and Hurd, R.G., Crop Processes in Controlled Environment, Academic Press, Oxford

Plant pathology and crop production

Burdon, J.J. and Leather, S.R., Pests, Pathogens and Plant Communites, Blackwell, Oxford, ISBN 0 6320 2561 1

Clifford, B.C. and Lester, E., Control of Plant Diseases, Costs and Benefits, Blackwell, Oxford, ISBN 0 6320 1453 9

Curr, S.J., McPherson, M.J. and Bowles, D.J., Molecular Plant Pathology, a practical approach (2 vols), Oxford University Press, Oxford, ISBN 0 1996 3353 3

Duncan, J.M. and Torrance, L., Techniques for the Rapid Detection of Plant Pathogens, Blackwell, Oxford, ISBN 0 6320 3066 6

Greenhalgh, R. and Roberts, T.R., Pesticides Science and Biotechnology, Blackwell, Oxford, ISBN 0 6320 1618 3

McKingley, R.G., Vegetable Crop Pests, MacMillan Press, London, ISBN 0 8493 7729 3

Palo, R.T., Plant Defences Against Mammalian Herbivory, CRC Press, New York, ISBN 0 8493 6550 3

Smith, I.M. *et al*, European Handbook of Plant Diseases, Blackwell, Oxford ISBN 0 6320 1222 6

The reader should be aware that numerous journals publish materials relevant to crop production. Some such as 'Plant Physiology and Plants' contain research reports on aspects of plants sciences. Others are aimed at agriculturalists and horticulturalists and include such journals as 'Acta Horticuluturae',' Horticultural Reviews', and 'The Journal of Horticultural Sciences'. Relevant government departments also produce pamphlets concerning specific issues such as the techniques available to identify or control particular plant pests/pathogens.

Index

A